Studies in Applied Mechanics 49

Introduction to Hydrocodes

Studies in Applied Mechanics

20. Micromechanics of Granular Materials (Satake and Jenkins, Editors)
21. Plasticity. Theory and Engineering Applications (Kaliszky)
22. Stability in the Dynamics of Metal Cutting (Chiriacescu)
23. Stress Analysis by Boundary Element Methods (Balas, Sládek and Sládek)
24. Advances in the Theory of Plates and Shells (Voyiadjis and Karamanlidis, Editors)
25. Convex Models of Uncertainty in Applied Mechanics (Ben-Haim and Elishakoff)
28. Foundations of Mechanics (Zorski, Editor)
29. Mechanics of Composite Materials - A Unified Micromechanical Approach (Aboudi)
31. Advances in Micromechanics of Granular Materials (Shen, Satake, Mehrabadi, Chang and Campbell, Editors)
32. New Advances in Computational Structural Mechanics (Ladevèze and Zienkiewicz, Editors)
33. Numerical Methods for Problems in Infinite Domains (Givoli)
34. Damage in Composite Materials (Voyiadjis, Editor)
35. Mechanics of Materials and Structures (Voyiadjis, Bank and Jacobs, Editors)
36. Advanced Theories of Hypoid Gears (Wang and Ghosh)
37A. Constitutive Equations for Engineering Materials
  Volume 1: Elasticity and Modeling (Chen and Saleeb)
37B. Constitutive Equations for Engineering Materials
  Volume 2: Plasticity and Modeling (Chen)
38. Problems of Technological Plasticity (Druyanov and Nepershin)
39. Probabilistic and Convex Modelling of Acoustically Excited Structures (Elishakoff, Lin and Zhu)
40. Stability of Structures by Finite Element Methods (Waszczyszyn, Cichoń and Radwańska)
41. Inelasticity and Micromechanics of Metal Matrix Composites (Voyiadjis and Ju, Editors)
42. Mechanics of Geomaterial Interfaces (Selvadurai and Boulon, Editors)
43. Materials Processing Defects (Ghosh and Predeleanu, Editors)
44. Damage and Interfacial Debonding in Composites (Voyiadjis and Allen, Editors)
45. Advanced Methods in Materials Processing Defects (Predeleanu and Gilormini, Editors)
46. Damage Mechanics in Engineering Materials (Voyiadjis, Ju and Chaboche, Editors)
47. Advances in Adaptive Computational Methods in Mechanics (Ladevèze and Oden, Editors)
48. Stability of Nonlinear Shells - On the Example of Spherical Shells (Shilkrut, Editor)

General Advisory Editor to this Series:
Professor Isaac Elishakoff, Center for Applied Stochastics Research, Department of Mechanical Engineering, Florida Atlantic University, Boca Raton, FL, U.S.A.
elishako@fau.edu

Studies in Applied Mechanics 49

Introduction to Hydrocodes

Jonas A. Zukas
Computational Mechanics Associates
Baltimore, USA

ELSEVIER

2004

Amsterdam - Boston - Heidelberg - London - New York - Oxford
Paris - San Diego - San Francisco - Singapore - Sydney - Tokyo

ELSEVIER B.V.
Sara Burgerhartstraat 25
P.O. Box 211, 1000 AE
Amsterdam, The Netherlands

ELSEVIER Inc.
525 B Street, Suite 1990
San Diego, CA 92101-4495
USA

ELSEVIER Ltd
The Boulevard, Langford Lane
Kidlington, Oxford OX5 1GB
UK

ELSEVIER Ltd
84 Theobalds Road
London WC1X 8RR
UK

© 2004 Elsevier Ltd. All rights reserved.

This work is protected under copyright by Elsevier Ltd, and the following terms and conditions apply to its use:

Photocopying
Single photocopies of single chapters may be made for personal use as allowed by national copyright laws. Permission of the Publisher and payment of a fee is required for all other photocopying, including multiple or systematic copying, copying for advertising or promotional purposes, resale, and all forms of document delivery. Special rates are available for educational institutions that wish to make photocopies for non-profit educational classroom use.

Permissions may be sought directly from Elsevier's Rights Department in Oxford, UK: phone: (+44) 1865 843830, fax: (+44) 1865 853333, e-mail: permissions@elsevier.com. Requests may also be completed on-line via the Elsevier homepage (http://www.elsevier.com/locate/permissions).

In the USA, users may clear permissions and make payments through the Copyright Clearance Center, Inc., 222 Rosewood Drive, Danvers, MA 01923, USA; phone: (+1) (978) 7508400, fax: (+1) (978) 7504744, and in the UK through the Copyright Licensing Agency Rapid Clearance Service (CLARCS), 90 Tottenham Court Road, London W1P 0LP, UK; phone: (+44) 20 7631 5555; fax: (+44) 20 7631 5500. Other countries may have a local reprographic rights agency for payments.

Derivative Works
Tables of contents may be reproduced for internal circulation, but permission of the Publisher is required for external resale or distribution of such material. Permission of the Publisher is required for all other derivative works, including compilations and translations.

Electronic Storage or Usage
Permission of the Publisher is required to store or use electronically any material contained in this work, including any chapter or part of a chapter.

Except as outlined above, no part of this work may be reproduced, stored in a retrieval system or transmitted in any form or by any means, electronic, mechanical, photocopying, recording or otherwise, without prior written permission of the Publisher. Address permissions requests to: Elsevier's Rights Department, at the fax and e-mail addresses noted above.

Notice
No responsibility is assumed by the Publisher for any injury and/or damage to persons or property as a matter of products liability, negligence or otherwise, or from any use or operation of any methods, products, instructions or ideas contained in the material herein. Because of rapid advances in the medical sciences, in particular, independent verification of diagnoses and drug dosages should be made.

First edition 2004

Library of Congress Cataloging in Publication Data
A catalog record is available from the Library of Congress.

British Library Cataloguing in Publication Data
A catalogue record is available from the British Library.

ISBN: 0-08-044348-6

∞ The paper used in this publication meets the requirements of ANSI/NISO Z39.48-1992 (Permanence of Paper). Printed in The Netherlands.

PREFACE

What is a hydrocode and where did it get that ridiculous name? Hydrocodes fall into the very large category of computational continuum mechanics. They were born in the late 1950's when, following the development of the particle-in-cell (PIC) method at Los Alamos National (then Scientific) Laboratory, Robert Bjork at the Rand Corporation applied PIC to the problem of steel impacting steel and aluminum impacting aluminum at velocities of 5.5, 20 and 72 km/s. This is cited in the literature as the first numerical investigation of an impact problem. Because such impact velocities produce pressures in the colliding materials exceeding their strength by several orders of magnitude, the calculations were performed assuming hydrodynamic behavior (material strength is not considered) in the materials. Hence, the origin of the term *hydrocode* – a computer program for the study of very fast, very intense loading on materials and structures. Today, the successors of PIC have considerably more material modeling capability and are applied to a very wide array of problems. Major efforts are under way to retrofit existing buildings against car bombs and other forms of explosive loading to safeguard against terrorist attacks. Standards of construction for nuclear reactor pressure vessels and other structures are being reviewed with the help of hydrocodes to include aircraft impacts. An extensive experimental–computational effort is taking place in the automotive industry worldwide to design crashworthy vehicles. The safety of people and equipment from failure of industrial components such as flywheels and the spread and containment of fragments from industrial accidents, are ongoing problems of concern not only to equipment designers but to industrial insurers as well. Hydrocodes are even used to design sports equipment and analyze the effects of dimples on golfballs, to make them go farther or fly straighter. Finally, with more and more man-made and natural debris orbiting the earth, there is the constant need to design space structures to withstand impacts from particles of various sizes. Even a dust particle moving at 20 km/s can penetrate a space suit or station, causing catastrophic evacuation. Almost every conventional area of study—aerospace, defense, chemical engineering, mechanical and civil engineering—needs to address problems of structural dynamics, material behavior and failure at high strain rates, fluid dynamics, wear and erosion with tools such as hydrocodes or their cousins, nonlinear structures codes and computational fluid dynamics codes. Such calculations are no longer performed in hydrodynamic mode yet the old name has stuck.

There are three approaches to problems involving the release of a large amount of energy over a very short period of time, e.g. explosions and impacts. Most of the work is **experimental** in nature. This is due to the fact that the problems are highly nonlinear and require information about material behavior at ultrahigh loading rates that is not generally available. Experiments can be ghastly expensive, costing up to millions of dollars to model full-scale explosions or aircraft impacts. Yet sometimes there is no other alternative to determine the cause of an event or to obtain information about preventing future disasters. **Analytical** approaches are possible if the geometries involved are relatively simple (plates, spheres, cylinders, etc) and the loading can be described through boundary conditions (such as a pressure distribution on parts of colliding objects), initial conditions (velocities, displacements, pressures) or a combination of the two. Here one begins with the governing equations of continuum dynamics and presses the analysis as far as possible. In order to be mathematically tractable, one-dimensional approaches need to be employed and frequently empirical input is employed in the form of a "material constant". Thus the application of analytical approaches is limited. They can, however, provide considerable insight at low cost. If programmed on modern desktop computers, hundreds of parametric studies can be performed in minutes. **Numerical** solutions are far more general in scope and remove any difficulties associated with geometry. As any higher order model, however, additional input is required to drive them. In this case, it is a description of material behavior at ultra-high loading rates, something not generally available save for a handful of materials. As all hydrocodes use an explicit method to advance solutions in time, very small time steps are required for stable solutions. This in turn requires long computational times even on the fastest computers at considerable cost. Still, as studies have shown, the judicious combination of numerical simulations and experiments can take months off the design time for design of prototypes and cut costs by millions.

The need for numerical approaches being clear, why a book on the subject now when hydrocodes have been around since 1960's? For one thing, hydrocodes are not black boxes. They require a knowledge of numerical methods in the code, to be sure, but, more importantly, they require a keen understanding of the physics of the problem being addresses. Some six months to two years is required to properly learn both the physics and its application to problems via codes. There is virtually no guidance for this. Learning occurs through contact with experienced scientists, preferably in a mentor relationship, a fair amount of trial and error, access to a database of relevant experiments and knowledge of material behavior at high strain rates. Fortunately there are now a number of books which address the physics of high pressure and high rate material behavior. However, there is nothing out there that would help the novice appreciate how much there is to be learned. In the past, Department of Energy organizations as well as private firms have offered resident and short courses in this field. In fact, the parts of the book dealing with fundamentals come from short courses I have taught on the subject. As far as I know, no such training is being offered currently.

This book is intended for professionals who are newcomers to the field of computations of high rate events and for those who must interact with them and use computational results in their work. The first part of the book recounts the relevant physics behind hydrocodes. It is valuable not only for that purpose but also for bringing together by way of reference the large body of literature, scattered through diverse journals, government and corporate reports and conference proceedings on the subject. The references alone can save someone new to the field six months of digging through the literature without guarantee of finding everything relevant. Two chapters are devoted to the mechanics of hydrocodes. The first covers all the basic ingredients – methods of discretization, kinematics (Lagrangian, Eulerian and hybrid approaches), models of material behavior and failure at high rates, artificial viscosity, time integration procedures and so on. The next chapter integrates these and shows the actual workings of a Lagrangian code. This is a unique approach I have never seen applied before and should answer many questions as to where exactly in a code do the various components fit. Because not all problems can be addressed in a Lagrangian framework, a chapter is devoted to alternatives, from simple Euler methods to the most current research in meshless methods. Copious examples are provided throughout to illustrate basic concepts. Chapter 7 covers the experimental methods which are used to generate high strain rate materials data. Stressed are their utility as well as limitations so a code user can appreciate not only what approach is needed for a particular problem but also how far it can be pushed.

To err is human, and Chapter 8 covers most of the major blunders that are made using hydrocodes. In their present state, their application is as much an art as a science. Practical suggestions regarding gridding, constitutive modeling and other aspects of computing are presented.

The book is not a collection of numerical formulas. Many texts exist on that topic. Instead, it is in the category of being a guidebook to a foreign subject. Although hydrocodes have been successfully used for quite some time, much mythology and confusion about them still persists. This book introduces the reader to the physics involved, the methods used to apply this to practical problems, the need for appropriate input for a given problem and the nonphysical results that can be obtained by failing to adhere to a few simple rules. The book is infused with much practical experience, both my own and that of colleagues, gathered in over 30 years of work in this area. As valuable as the text are the cited references. The combination will take years off the preparation time of a code user.

In short, it is the book I would have liked to have had when I first started work in this field more years ago than I care to remember.

Jonas A. Zukas
Baltimore, MD

CONTENTS

Preface .... *v*

Chapter 1
Dynamic Behavior of Materials and Structures

1.1 Introduction 1
1.2 Structural dynamics problems 4
1.3 Wave propagation problems 16
1.4 Mixed problems 25
1.5 Summary 26
References 27
Further reading 30

Chapter 2
Wave Propagation and Impact

2.1 Introduction 33
2.2 Wave propagation in rods and plates 36
    a) Describe the wave motion set up in the body. 36
    b) Describe the wave motion set up in the striker. 41
2.3 Flies in the ointment 49
2.4 Bending waves 55
2.5 Wave reflections and interfaces 55
    a) Discontinuous cross-sections and different materials. 57
    b) Wave propagation involving discontinuous cross-sections. 59
    c) Layered media. 60
    d) Oblique impact. 66
2.6 Dynamic fracture 66
2.7 Summary 70
References 72

Chapter 3
Shock Waves in Solids

3.1 Introduction 75
3.2 Uniaxial strain 76

x   Contents

3.3 Wave propagation 82
3.4 Conservation equations under shock loading:
     The Rankine-Hugoniot jump conditions 85
3.5 The Hugoniot 87
     a) The U-u plane. 88
     b) The P-V plane. 89
     c) The P-u plane. 91
3.6 The equation of state 91
     a) Mie-Gruneisen. 94
     b) Tillotson. 95
     c) Explosives. 98
3.7 Summary 99
References 101

Chapter 4
Introduction to Numerical Modeling of Fast, Transient Phenomena

4.1 Introduction 103
4.2 Spatial discretization 105
     a) Finite differences. 111
     b) Finite elements. 117
4.3 Lagrangian mesh descriptions 123
     a) Mesh characteristics. 123
     b) Contact-impact. 126
     c) Large distortions. 134
4.4 Artificial viscosity 138
4.5 Time integration 141
4.6 Constitutive models 148
     a) Constitutive descriptions for metallic materials. 150
     b) Constitutive descriptions for non-metallic materials. 155
4.7 Problem areas 160
4.8 Summary 161
References 162

Chapter 5
How Does a Hydrocode Really Work?

5.1 Introduction 169
5.2 Pre-Processing 169

5.3 Number crunching 174
   a) Velocities and displacements. 174
   b) Contact. 176
   c) Strains and strain rates. 185
   d) Stress, failure and energy for inert materials. 186
   e) Pressure and energy for explosive materials. 193
   f) Nodal forces. 195
5.4 Once more, with feeling 198
References 198

Chapter 6
Alternatives to Purely Lagrangian Computations

6.1 Introduction 201
6.2 Euler codes 204
   a) Material interfaces and transport. 210
6.3 Arbitrary Lagrange–Euler methods (ALE) 223
6.4 Coupled Euler–Lagrange calculations 226
6.5 Smoothed-particle hydrodynamics (SPH) 235
References 246

Chapter 7
Experimental Methods for Material Behavior at High Strain Rates

7.1 Introduction 251
7.2 The split-Hopkinson pressure bar (Kolsky apparatus) 252
7.3 The Taylor cylinder 260
7.4 The expanding ring 265
7.5 Plate impact experiments 268
7.6 Pressure–shear experiments 272
7.7 Summary 275
References 276

Chapter 8
Practical Aspects of Numerical Simulation of Dynamic Events

8.1 Introduction 279
8.2 Difficulties 281

8.3	Idealization	283
8.4	The human factor	284
8.5	Problems related to computational meshes	285
	a) Element aspect ratio.	285
	b) Element arrangement.	287
	c) Uniform and variable meshes.	289
	d) Abrupt changes in meshes.	296
8.6	Shortcuts	297
8.7	Summary	306
References		307
Index		311

Chapter 1
Dynamic Behavior of Materials and Structures

1.1 Introduction.

All processes in life are dynamic.

This may sound drastic. Think about it, however. Can you name a single process that does not depend, in one fashion or another, on time? I cannot, although I can think of several where the time involved observing a physical change is so large that, for all practical purposes, we can treat the process as being independent of time.

Figure 1.1, due to Lindholm [1971], is a good illustration of this point. On the left are processes that involve such a large period (or alternatively, such a low rate of change of the phenomenon with time) that we think of them as static (time-independent) processes. As we move to the right, the time frame decreases (the rate of change with time increases) and other phenomena (such as inertia) begin to play a role in the solution of the problem. On the right, we deal with events that occur in very short intervals of time. Vehicle collisions, impacts of space debris with orbiting objects, explosions, lunar and planetary impacts are but a few examples here. With decreasing time comes increasing complexity. Direct measurements of key parameters such as stress, strain, temperature and pressure become difficult or impossible. They have to be inferred from indirect measurements. This, however, requires some knowledge of the material behavior at these very high strain rates. Quite often, it is a model, or constitutive relationship, which we seek. Thus, we are caught in a quandary. We experiment to model the high rate behavior of materials, but we need such a model to interpret the experiments!

It might help our understanding of dynamic phenomena in mechanics a bit if we divide the continuum of dynamics problems into two groups:

- Structural dynamics.

- Wave propagation.

There is no clear demarcation between these two areas. Indeed, the labels are misleading since both types of problems involve wave propagation. Nonetheless, these designations have caught on in the literature, so we will make the best use of them that

2 *Introduction to Hydrocodes*

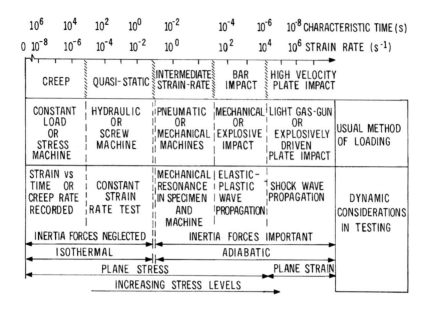

Figure 1.1 Material behavior with changing strain rate and load duration [Lindholm, 1971].

we can. These two labels deal with the behavior of inert (non-explosive) materials, which are subjected to

- intense impulsive loads [distributed over a surface, such as air blast over a fairly long time (milliseconds to seconds)] or

- impact [applied to a single point or a very small area over a very short (nanoseconds to microseconds) time span].

There is also a large class of energetic materials that react quite differently when excited. To cover these, we will add a third category, *Explosives*. Energetic materials will not be discussed in this book. For information on this topic, see Zukas & Walters [1997], Cooper & Kurowski [1997], Cooper [1996], Person, Holmberg & Lee [1994], Cheret [1993], Fickett & Davis [1979] and Mader [1979, 1997].

The same equation governs phenomena in both categories. It is most easily recognized in its one-dimensional (1D) form

$$c^2 \frac{\partial^2 u}{\partial x^2} - \frac{\partial^2 u}{\partial t^2} = f\left(u, \frac{\partial u}{\partial t}, x, t\right). \tag{1.1}$$

This is the familiar *wave equation*. Here, u represents displacement; f a generalized loading function, which depends on displacement, velocity and spatial (x) or temporal (t) quantities. If you are not familiar with the wave equation, there are a number of excellent introductory texts on the subject. The books by Bedford and Drumheller [1993], Main [1993] and Drumheller [1998] are both comprehensive and eminently readable. The first develops the necessary mathematical background required to deal with linear elastic waves in solids. The text by Main presents a wealth of applications of the wave equation to various problems in physics, including free and forced vibrations, dispersive and non-dispersive waves, water waves, electromagnetic waves, De Broglie waves and solitary waves. Drumheller's most recent book is a lucid compendium of information on nonlinear waves, constitutive models and equations of state. It is especially valuable to those working in computational approaches to problems in dynamics.

The general solution of the wave equation can be obtained in one of two forms: the *normal mode* solution or the *traveling wave* solution. The normal mode solution is a technique for solving structural dynamics problems in terms of variables that are multiples of the free, undamped mode shapes (eigenvectors) of the structure. The solution is postulated to be separable into functions of the spatial and temporal variables, that is

$$u(x,t) = \Phi(x)K(t). \tag{1.2}$$

An eigenvalue problem for the structure is formed, the appropriate eigenvalues and eigenvectors determined, and the results written in the form

$$u = \sum A_i \sin(\omega_i t)\sin \frac{\omega_i x}{c}, \tag{1.3}$$

where the ω_i are the eigenvalues of the system. The summation is carried out to the desired degree of accuracy. This approach, usually called *modal analysis*, is useful in situations where the response of the structure is affected principally by a few dominant modes, usually the lower harmonics. In the past, structural dynamics inevitably meant problems associated with shock and vibration situations. The linear response of a structure could be easily handled with a normal mode solution. The technique was suitable to vibration problems and seismic loading situations, as well as related situations. Today, the term structural dynamics has a much broader meaning.

The second type of solution to the wave equation is known as the *traveling wave approach*. Here, the disturbances induced by the applied loads are viewed as propagating waves, which can interact with geometric and material boundaries as well as each other. It can be shown [Achenbach, 1975; Wasley, 1973] that the most general solution of the 1D wave equation is

$$\psi = f(x - ct) + g(x + ct), \tag{1.4}$$

where f and g are arbitrary functions of the arguments, $x - ct$ and $x + ct$, respectively. The quantities f and g must be consistent with the requirements for continuity, small amplitude, and the various imposed boundary conditions. Two functions are necessary because the wave equation is of second order.

This approach to the solution of the wave equation is most appropriate (and computationally efficient) for problems involving intense, short-duration loading, especially where the energy carried by waves can result in material failure when the waves interact with material interfaces, free surfaces, or each other. Examples include lunar and planetary impacts, the impact of meteoroids and assorted space debris with space structures at velocities of 5–20 km/s as well as the penetration and perforation of solids impacted by high-speed missiles, high-obliquity ricochet, the erosion and fracture of solids subjected to impacts by kinetic energy projectiles or shaped charge jets and the explosive loading of metals.

1.2   Structural dynamics problems.

A critical problem today is the response of vehicles and structures to internal or external *blast loading*. The source of the blast can be detonation of a high explosive, a fuel–air or vapor cloud explosion, particulate or dust explosions, pressure vessel failures or boiling liquid expanding vapor explosions—the flash boiling of liquid due to sudden exposure to an intense heat source. Some examples are shown in Figs. 1.2 and 1.3.

There exists a vast literature on this subject, dealing with both free-field effects (e.g., see Kinney & Graham [1985] and Baker [1973]) and the interaction of a blast wave with a structure [e.g., Baker et al., 1983]. The blast and thermal aspects of nuclear weapons effects have been studied for some time [e.g., Glasstone & Dolan, 1977]. Problems in this category cannot be solved from first principles. Instead, graphical depictions of pressure–time histories for confined and unconfined explosions exist, together with a host of empirical data. This information is used to describe a surface load as a function of time to a finite-element/finite-difference computer code, which can then grind out the damage expected in a structure. Except for simple geometries, such calculations are very expensive as they require rather fine resolution to capture all of the possible deformation modes. Experience is required, to both set up and interpret calculations. Computer codes based on empirical input and empirical equations are also available. Needless to say, these require even greater caution in their application to

Figure 1.2 Detonation of two 1000 lb spheres of high explosives stacked vertically (Courtesy: Dr. C. Weickert, DRES).

Figure 1.3 Detonation of a fuel–air explosive line charge (Courtesy: Dr. C. Weickert, DRES).

design problems. Current interest is in problems involving terrorist attack and failure of industrial equipment. The initial blast loading can give rise to free-flying debris of various shapes and densities, causing further damage to nearby structures or civilians through impact and penetration. Current efforts to protect civilian structures from blast loads deal with problems of missile penetration, determination of collision forces on structures, establishment of the mechanics of failure under dynamic loads and related topics can be found in various conference proceedings [e.g., Jones, Brebbia & Watson, 1996; Brebbia & Sanchez-Galvez, 1994; Bulson, 1994; Geers & Shin, 1991; Rajapakse & Vinson, 1995; Shin, 1995; Shin & Zukas, 1996; Shin, Zukas, Levine & Jerome, 1997; Levine, Zukas, Jerome & Shin, 1998]. Problems involving structural impact are dealt with by Jones [1989], Johnson [1972] and Macaulay [1987], among others.

Another major area of concern is *fluid–structure interaction*. This is a compact title, which hides a wide variety of applications. These can include interaction of offshore structures with water waves, especially during violent storms, sloshing of liquids in open and closed containers, the action of water against dam-like structures, hydrodynamic ram effects (Fig. 1.4), bursting of fluid-filled containers, vibration of structures in fluids, interaction of waves with flexible structures, excitation of turbomachine blades, flutter and flutter control and many others. The problem involves the interaction of a low-density, highly deformable medium with one that is more dense and orders of magnitude stiffer. Computationally, the problem is addressed through codes that have Eulerian features or use a smooth-particle hydrodynamics approach (meshless codes) to cope with the severe deformation gradients. Problems involving *liquid–solid impacts* can be included in this category or be listed separately. They are of interest because they can easily give rise to stress waves that can cause large deformation and failure in impacted materials. Forward regions of high-speed aircraft can be damaged on passage through a raincloud. The type and extent of damage that occurs in liquid–solid impact depends primarily on the size, density, and velocity of the liquid and on the strength of the solid. At a velocity of 750 m/s, a 2 mm diameter water drop is capable of fracturing and eroding tungsten carbide and plastically deforming martensitic steels. At velocities near 200 m/s, a single impact may produce no visible damage but repeated impacts will bring about erosion of the solid. The resulting weight loss in the solid does not occur at a constant rate but can be divided into several stages (see Brunton & Rochester [1979] for a thorough discussion). There is an initial incubation period during which there is no detectable weight loss, although some surface deformation may be noted. This is followed by a period of increasing material removal as cracks and voids coalesce. Finally, the erosion rate levels off and the weight loss continues at an approximately constant rate. Additional information on liquid–solid impacts may be found in the works of Springer [1976], Preece [1979] and Adler [1979]. *Jet cutting technology* also can be included in

Dynamic Behavior of Materials and Structures 7

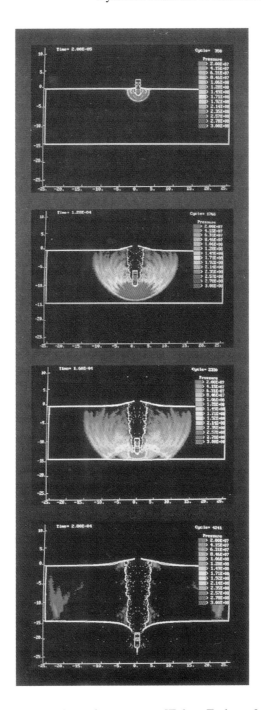

Figure 1.4 Stages of hydrodynamic ram event [Zukas, Furlong & Segletes, 1992].

this problem area. Water jets are used in rock cutting for mining as well as industrial applications for making furniture, puzzles and shoes, among other things. Finally, there is the always fun problem of *water hammer*. Fluid transients result from disturbances, either from within a system or without, which create pressure waves. These waves may be either pressure increases or decreases, which travel through a system. The result of such pressure waves is varied and depends greatly on the magnitude of the pressure waves and the system within which these waves travel. As a minimum, the pressure waves cause localized stressing of the system and in severe cases can result in failure of the system boundary or damage and possible failure to adjacent components resulting from gross system movement.

The pressure waves can be caused by many conditions such as the collapse of a void, slug flow against an area restriction or pipe bend, a pump start, control or relief valve actuation, water suspended in a steam line or other such events which are of a transient nature to the system. In each case, however, it is the physical interaction of fluid against fluid or fluid against a structure which causes the pressure waves and the magnitude of these pressure waves can be related directly to the amount of energy the fluid must gain or lose [Watkins & Berry, 1979].

Another broad grouping which encompasses a number of important technological problems is *mechanical system dynamics*. Included here are problems associated with:

General machinery and mechanism systems. A partial listing of systems would include production equipment, material handling systems, bulk and unit conveying equipment, machine tools, engines, robots, and many more.

Agricultural, construction and off-highway systems such as farm tractors, construction equipment, fork lift trucks, utility vehicles and so on.

Turbo machinery systems, such as power generation equipment, turbines, compressors, pumps and their drive systems.

Containment structures for failed components of turbo machinery systems, particularly aircraft engines and flywheels, to mention but a few. The article by Ashley [1996] summarizes current problems and solutions. In a related area, design of containment structures for nuclear reactors to guard against internal pressurization from reactor excursions or impact by debris resulting from failed components as well as external loads from tornado-borne debris and aircraft impacts is an ongoing problem. Also crucial is the design of shipping casks for the safe transportation of hazardous materials. For a good example of current capabilities, see the article by

Dynamic Behavior of Materials and Structures 9

Figure 1.5 Simulation of car crash using DYNA. The computational model was developed by FHWA/NHTSA National Crash Analysis Center, George Washington University.

Morandin & Nadeau [1996]. The article by Key [1985] shows some of the difficulties in computing deceptively simple geometries.

Vehicular collisions. Attention has been focused of late on automotive crashworthiness problems [see the Fall, 1996 issue of *IRIS Universe* on the use of simulation in auto design]. Much of the software developed for defense applications (codes such as DYNA and NIKE) and various derivatives such as PAM-CRASH are now being applied to automotive impact situations (Fig. 1.5). BMW saves an estimated $1 million for every real crash that is replaced with a computer simulation. Also of interest are train and ship collisions. Background information can be found in the book by Johnson & Mamalis [1978].

Aeronautical and aerospace systems such as landing gears, helicopter rotors, gyros and solar panel deployment systems and luggage containment systems to mitigate explosions.

Many more could be listed but these will suffice to make the point that there is no lack of application for existing technology and no lack of demand for advancing that technology. Models at all levels are needed to reduce costs, reduce product development time and enhance the extraction of information from experiments. In all these diverse problems, it is necessary to simulate

- ranges of motion of system components;

- transient response during startup;

- loading on parts;

- resonances, mode shapes, transfer of forces and torques;

- bending of parts;

- overshoots;

- timing analysis.

In addition, models are needed to troubleshoot existing systems and help design new ones.

The study of *biodynamical systems* has advanced to the point of including finite elements in the arsenal of simulation tools. Originally, rigid-body dynamics was used in simulations. Various vehicle components, as well as occupants, were modeled as an assemblage of rigid bodies connected by nonlinear springs. These simple models were able to give some indication of the loads experienced by vehicle occupants in frontal and side impacts, as well as the displacements to be expected by steering columns, doors, rocker panels and the auto body. Various sub-models for body dynamics, contact, restraint systems and injury criteria could be combined in a calculation. Currently, finite elements are used to model automotive crashes as well as occupant body deformation. Such simulations require more than just raw computing power. With a refinement of the model to include deformable mechanics comes a requirement for material properties and material constitutive models at the appropriate strain rates encountered in a vehicle collision. Calculations with models and data appropriate under quasi-static conditions produce very expensive garbage when applied to transient situations. A large database of dynamic material data exists for metals, although the information is scattered throughout the literature and requires some ingenuity in retrieval. For composite materials and for the components of the human body, however, neither high-rate models nor data exist.

This, rather than any lack of computing power, will serve as the greatest limitation on the growth of this area for the next decade.

Problems involving *beam, plate and shell* structures are among the most common in practice. Computationally, they are the most difficult to solve. It has been said, in fact, that 3D problems are more easily solved numerically than problems involving shell structures [Steele, Tolomeo & Zetes, 1995]. The complexities arise from dealing with curved geometry and bending. This is an area of considerable and intense research and some excellent computational results have been achieved. Users of computer codes which include plate and shell elements must have a thorough understanding of the theory and some expectation of what the computed results will be. Interpretation of code results is a non-trivial exercise. An example of a dynamic nonlinear deformation of a shell-like structure with the DYNA code is shown in Fig. 1.6.

Non-perforating impacts are of interest because of the stress waves generated by the impact. These will propagate in the impacted body, even though the striker is arrested. Once the propagating wave encounters a free surface or a material interface, it will be partially transmitted and partially reflected. Tensile waves can be generated which can

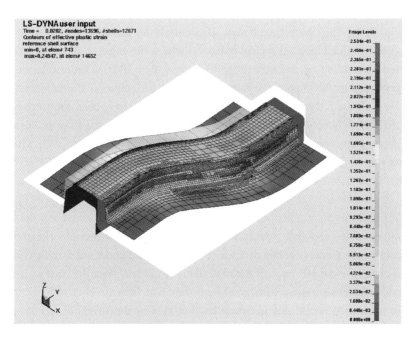

Figure 1.6 Large deformation of a shell-like structure using LS-DYNA. The computational model was developed by FHWA/NHTSA National Crash Analysis Center, George Washington University

cause material failure or damage critical components within. Extensive tests for soft-body (bird) impacts against aircraft have been documented by Wilbeck [1978]. Shock spectra generated by impacts against tank armor have been reported by Quigley [1989, 1991]. These pressure profiles and peak pressure magnitudes can be used in subsequent studies to assess damage.

Many more examples of structural dynamics situations could be cited here. For example, we have made no mention of problems associated with *rotating machinery, metal forming, underwater explosions,* and problems associated with materials. These, and many others, are covered in a valuable book by Pilkey & Pilkey [1995], which reviews computational methods and computer programs for a wide variety of structural dynamics problems. As diverse as these problems may seem, they have several unifying characteristics:

(a) Structural dynamics problems typically involve loading and response times in a time frame of *milliseconds to seconds.*

To get an appreciation for this, consider the collision of two automobiles. At typical automotive speeds, it would take several seconds for the exterior damage to reach its terminal state and for the loads to be transferred to the occupants of the vehicle. The stresses involved may be large and the extent of damage fearful, but it would take several seconds for the vehicles to deform and more seconds as they decelerate to a halt, moving as though they were rigid bodies. By contrast, let us say a bullet is fired into an automobile's engine block. At typical bullet velocities, the engine will be penetrated (or perforated, depending on the bullet's path) in less than 100 μs. From then on, stress waves generated by the impact will propagate through the automotive structure until they are finally dissipated. By about 1 ms after impact, these will have been almost completely dissipated. Thus, structural dynamics problems and wave propagation problems (the bullet impact) are separated in time by several orders of magnitude and, with some exceptions to be mentioned later, may be treated in an uncoupled fashion.

(b) Typical strain rates for structural dynamics problems are in the range of 10^{-2}–10^2 per second.

This gives us some idea of the dynamic response to be expected and the type of constitutive model and data which we need to model these events. Handbooks of structural alloy properties contain data which was generated under quasi-static conditions at strain rates of about 10^{-4} s^{-1}. If the material is *strain rate sensitive*, that is, if there is considerable increase in stress with increasing strain rate, then use of quasi-static data in a dynamic simulation is not only inappropriate but may be downright dangerous since it will produce erroneous results.

It is difficult to overemphasize the importance of strain rate since very expensive calculations using inappropriate data are so often performed. If the material is strain rate sensitive, then this must be accounted for in analyses or numerical simulations to avoid

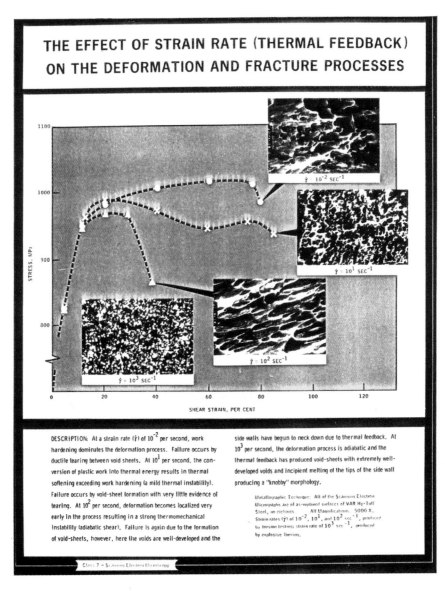

Figure 1.7 Changes in load-carrying capability and failure mode with increasing shear strain rate (Courtesy: Dr. C. Hargreaves).

generating expensive garbage. Consider Fig. 1.7 above. It shows the behavior of a high-strength steel at various shear strain rates ranging from 10^{-2} to 10^3 s^{-1}. Not only does the load-bearing capability change, as shown by the different shear stress–strain curves, but there are also significant differences in the material's failure mechanism with increasing strain rate. At the lowest strain rate, work-hardening dominates the behavior and failure occurs by ductile tearing between void sheets. As strain rate increases the conversion of plastic work into thermal energy results in thermal softening exceeding work-hardening. Failure occurs by void sheet formation with little evidence of tearing. At a shear strain rate of 10^2 s^{-1}, the deformation becomes very localized in the process resulting in a strong thermomechanical instability known as adiabatic shear. Failure is again due to the formation of void sheets but here the voids are well developed and the sidewalls have begun to neck down due to the thermal feedback. At 10^3 the deformation process is adiabatic and the thermal feedback has produced void sheets with extremely well-developed voids and incipient melting of the tips of the sidewall.

(c) Typical strains for structural dynamics problems are on the order of 0.5–10%.

Most plasticity theories in use for structural problems are valid for strains between 3 and 10% because of the assumptions inherent in their derivation. This is fine for structural dynamics situations since similar degrees of plastic deformation are obtained in reality. It is inappropriate for wave propagation calculations where strains on the order of 60% or more are encountered, as are pressures which exceed material strengths by an order of magnitude.

(d) The magnitude of the hydrodynamic pressure generated is *on the order of the material strength*.

In structural dynamics calculations the pressure, if it is ever needed, is computed from the normal stresses as

$$p = \frac{1}{3}(\sigma_{11} + \sigma_{22} + \sigma_{33}). \tag{1.5}$$

By contrast, in wave propagation problems, the pressure can exceed material strength by orders of magnitude.

(e) Structural dynamics problems involve *global* deformations caused primarily by the *lowest harmonics* of the system. These come about from the hundreds or thousands of wave transits which occur in the structure. Figure 1.8 shows the difference between local and global response. Another example would be a car crash. The whole car and its occupants respond in a crash. An entire building, or a very significant portion of it, is affected by a terrorist car bomb. As a rule, we cannot isolate a single beam or wall or other component and be assured that the entire structure will behave in a similar manner.

(f) Material or structural failure is, in many cases, inferred from the global deformations. If a structure must remain elastic, then the presence of plasticity constitutes failure in the sense of an unacceptable design. If plastic deformation is acceptable, it may be limited to a few percent or allowed only in certain portions of

LOCAL - GLOBAL RESPONSE

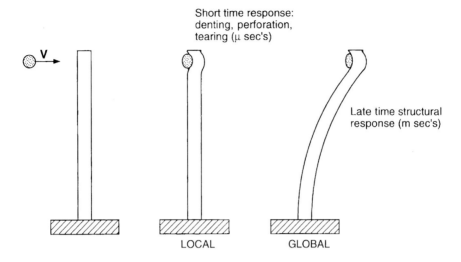

Figure 1.8 Local vs. global response to impact.

the structure. As a rule, there is no *a priori* specification of a damage function which is incorporated into the calculation and affects results.

The computer codes which have been developed for structural dynamics problems reflect the physics governing dynamic response. Because the deformations are limited and global, a Lagrangian description (where the computational grid is embedded in the material and deforms with it) is favored. If the distortions become too large, an arbitrary Eulerian–Lagrangian (ALE) approach may be used for problems that fall between the structural dynamics and wave propagation categories. Here, some portion of the computational grid is allowed to move to accommodate local velocity gradients. No definitive kinematic model has emerged, although many codes employ a small-strain, large rotation formulation.

Initially in structural codes (MARC, ADINA, ANSYS, ABAQUS, STARDYNE, NIKE, to name but a few) an implicit temporal integration scheme (Newmark—β, Houbolt, Wilson—θ) was employed to advance the solution in time. The newest codes employ both a wider catalog of implicit schemes as well as explicit methods. These have been found to be fast and accurate for problems involving metal forming and vehicle crashworthiness studies. Indeed DYNA, an explicit code originally developed for high velocity impact problems, has become a mainstay in the crashworthiness arena and many of the codes used abroad are direct descendants of DYNA. The material data required for structural dynamics calculations can often be obtained from various mechanical testing machines and wave propagation experiments (see Meyers [1994] and Nicholas & Rajendran [1990] for discussions of high rate material test techniques).

Structural dynamics codes contain a large element library, including various types of beams, plates, shells, rods, connectors, contact elements and 2D and 3D solid elements to allow flexibility in modeling the various structures and loading conditions encountered in reality. Higher-order elements are incorporated to complement linear and constant strain elements. Along with this, a host of surface and body loading options, various types of boundary conditions and initial conditions are available. The early structural dynamics codes made no provision for contact–impact problems. A contact had to be simulated as some form of surface loading. Results, it turned out, were very sensitive to the assumed loading functions. Current codes incorporate algorithms to allow explicit treatment of contact problems.

1.3   Wave propagation problems.

Problems involving *lunar and planetary impact* have been of interest since the early 1600s and are closely related to the invention of the telescope and the discovery of craters on the moon. The fundamentals of impact cratering in geological materials are described in the excellent book by Melosh [1989]. The impacts of asteroids, meteorites and comets on planetary surfaces produce craters and eject mass into space. The surface of the moon is a sequence of craters of different sizes. The impact of space matter on the earth's surface has given rise to interesting speculations. One of the more popular ones is the theory for the extinction of dinosaurs due to a large meteorite impact. Another is that the entire Everglades area in southern Florida could have arisen from a large crater. The impact on Jupiter by the fragmented comet Shoemaker–Levy is another example of continuing dynamic events in space.

Explosive welding is an important example of a successful commercial application of explosives to materials. It is a valuable adjunct to conventional welding for metal combinations such as titanium–steel, copper–aluminum and copper–steel which cannot be welded in conventional ways. Reviews have been written by Rinehart & Pearson [1963], Deribas [1972], Crossland [1982] and Blazynski [1983], among others. In the basic setup, an inclined impact between two metallic objects creates a jet that cleanses the surfaces of all contaminants and forces them into direct contact. The pressure due to the explosive acts upon these two clean and hot surfaces for a time sufficient for bonding and cooling to occur. The bonding is a result of shear localization at the surface. The welded interface often has a wavy configuration. Examples are to be found in the books by Meyers [1994] and Blazynski [1983].

Explosive forming uses the high gas pressures created by the detonation of explosives to form metals. It has an advantage over conventional forming methods when small batches of large pieces have to be produced because investment in dyes is minimized. It is also useful in situations where a metal exhibits an enhanced formability at high strain rates. Explosive forming can be accomplished in air or water [Meyers, 1994].

Explosive hardening of metals is the result of passing a high-amplitude shock wave through a metal in a state of uniaxial strain. This produces plastic deformation without a change in shape. The most successful industrial application of shock hardening has been for Hadfield steel. This is a high-manganese austenitic steel with high strength, toughness, and wear resistance. By shock hardening the surfaces, the wear resistance is tripled [Meyers, 1994].

Shock consolidation and *shock synthesis* have also been extensively studied but with few commercial successes. See Murr [1990] and Batsanov [1993] for discussions and applications. The area of *blasting* is more concerned with explosives than with high strain rate behavior of rock and geological materials. See Persson, Holmberg & Lee [1993] for more on this subject.

Phenomena involving penetration, perforation and ricochet have been studied extensively for over two centuries. This area can be divided into problems involving *kinetic energy penetrators* and *chemical energy penetrators*.

Chemical energy penetrators, also known as shaped charges, Miznay–Chardin devices, self-forging fragments, detonating rounds and a variety of other names derive their energy from the detonation of an explosive when the system approaches or is in direct contact with a target. The basic geometry consists of a cylinder of explosive with a hollow cavity at one end and a detonator at the opposite end. The hollow cavity, which may assume almost any geometric shape (hemisphere, cone, etc.) causes the gaseous products formed from the initiation of the explosive at the end of the cylinder opposite the hollow cavity to focus the energy of the detonation products. This focusing creates an intense, localized force. When directed against a metal plate, this concentrated force is capable of creating a deeper cavity than a cylinder of explosive without a hollow cavity, even though more explosive is available in the latter case.

If the hollow cavity is lined with a thin layer of metal, glass, ceramic or the like, the liner forms a jet when the explosive charge is detonated. Upon initiation, a spherical wave propagates outward from the point of initiation. This high-pressure shock wave moves at a very high velocity, typically around 8 km/s. As the detonation wave engulfs the lined cavity, the material is accelerated under the high detonation pressure, collapsing the cone. During this process, depicted in Fig. 1.9 for a typical conical liner, the liner material is driven to very violent distortions over very short time intervals, at strain rates of 10^4–$10^7\,\text{s}^{-1}$. Maximum strains greater than 10 can readily be achieved since

18 *Introduction to Hydrocodes*

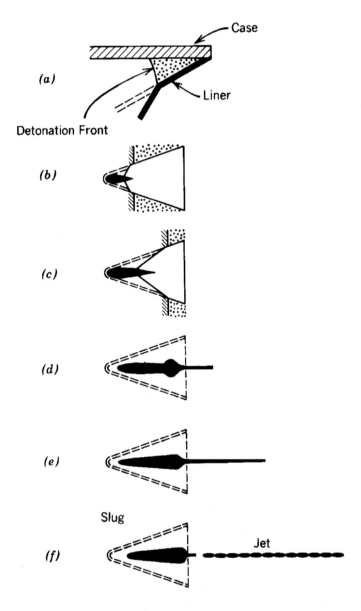

Figure 1.9 Conical shaped charge liner collapse and jet formation [Walters & Zukas, 1989].

superimposed on the deformations are very large hydrodynamic pressures. In fact, the peak pressure is approximately 200 GPa, which decays to an average of approximately 20 GPa. The collapse of the conical liner material on the centerline forces a portion of the liner to flow in the form of a jet where the jet tip velocity can travel in excess of 10 km/s.

Because of the presence of a velocity gradient, the jet will stretch until it fractures into a column of particles.

When this extremely energetic jet strikes a metal plate, a deep cavity is formed, exceeding that caused by a hollow charge without a liner. Peak pressures in the metal plate of 100–200 GPa are generated, decaying to an average of 10–20 GPa. Average temperatures of 20–50% of the melt temperature and average strains of 0.1–0.5 are common. Localized temperatures and strains at the jet tip can be even higher. The penetration process occurs at strain rates of 10^6–10^7 s^{-1}. The cavity produced in the metal plate due to this jet–target interaction is due not so much to a thermal effect but to the lateral displacement of armor by the tremendous pressures created (Fig. 1.10). The target material is actually pushed aside and the penetration is accompanied by no change in target mass, neglecting any impact ejecta or spall from the rear surface of the target.

The many aspects of chemical energy projectile formation, breakup, penetration and application are covered by Walters & Zukas [1989], Walters [1990] and Carleone [1993].

Kinetic energy penetration phenomena are of interest in a variety of problem areas. Though often associated with military applications (terminal ballistics), the same considerations apply in problems involving containment of high mass or high velocity debris due to accidents or high-rate energy release, safety of nuclear reactor containment vessels, design of lightweight body armors, erosion due to multiple high-speed impacts and protection of spacecraft from meteoroid impact. A kinetic energy projectile uses the energy of its motion to penetrate and perhaps perforate (completely pierce) a barrier. At high obliquity impacts, it may also ricochet (bounce off) from the impacted surface, Figs. 1.11 and 1.12.

Upon impact, compressive stress ways are generated in both striker and target. They move with either the sound speed of the materials making up the striker and target for subsonic impacts or at the shock velocity for hypervelocity impacts (impacts at velocities exceeding the speed of sound in a material—usually ≥ 6 km/s in metals). These are followed by slower moving shear waves. For sufficiently high impact velocities, relief waves will be generated in the rod due to the presence of lateral free surfaces, creating a 2D stress state behind the compressive front for normal impacts but 3D ones for oblique impacts, since now asymmetric bending waves are involved. In the target, the two wave systems will propagate until they interact with a material interface or a free surface. There, in order to satisfy boundary conditions, tensile waves can be generated. If *both* the *amplitude* of the tensile stress pulse and its *duration* at a point in the material are sufficient, material failure by a variety of mechanisms can occur.

There are two basic problems in penetration mechanics. Given the geometry, initial conditions, a well-characterized material:

20 *Introduction to Hydrocodes*

Figure 1.10 Jet penetration of armor [Walters & Zukas, 1989].

(1) Determine the depth of penetration if the target is semi-infinite.

(2) Determine the residual mass and velocity of the projectile and the hole characteristics in the target if it is of finite thickness.

Dynamic Behavior of Materials and Structures 21

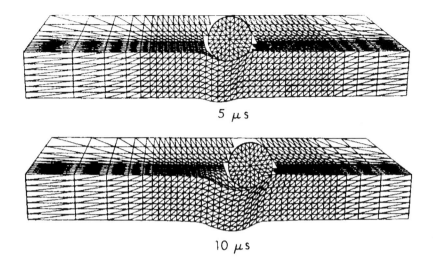

Figure 1.11 Early stages of steel sphere impact on steel target at 75 degrees obliquity (measured from the target normal) [Zukas *et al.*, 1982].

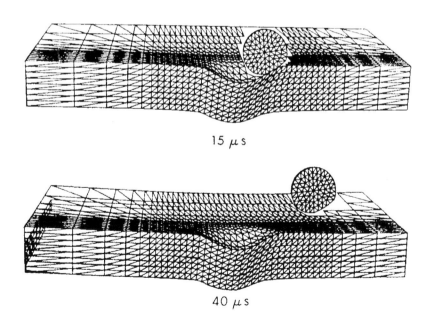

Figure 1.12 Cratering and liftoff [Zukas *et al.*, 1982].

Because of the complexities in oblique impact due to high rate material behavior, material failure due to stress wave interactions, the geometries of the problem and the presence of bending waves, closed form solutions are rarely, if ever, available. Therefore most of the work in this area is experimental in nature.

Large databases exist, as do analytical models, to allow decent guesstimation of maximum penetration depth. Models also exist to estimate hole size. Both problems readily lend themselves to numerical computations, though these are still expensive, if routine. As a last recourse a few experiments can always be performed.

The fundamental problem of impact against a finite thickness barrier or a thin-walled structure is the determination of a ballistic limit—a velocity below which the structure will not be perforated. Many thin-walled structures are unable to stop energetic projectiles or fragments. Then it becomes necessary to know the residual mass and velocity of the projectile after perforation. Ballistic limits are discussed extensively in Chapter 5 of Zukas, Nicholas, Swift, Greszczuk & Curran [1982]. Hypervelocity aspects of ballistic limit determinations can be found in the proceedings of the various Hypervelocity Impact Symposia (1987–1995). If the striker has a simple geometric form, e.g., right circular cylinder, sphere, cone, then a number of analytical and empirical relationships can be used to determine the ballistic limit and residual projectile characteristics. These can be found in the books by Zukas *et al.* [1982, 1990], Walters & Zukas [1989], Macaulay [1987], Johnson [1972] and Backman [1976], as well as the comprehensive survey of penetration mechanics by Backman & Goldsmith [1978]. When the striker has a non-standard or irregular geometry, or when the target is complex and consists of more than a single plate element, the ballistic limit must be determined either experimentally or through recourse to wave propagation computer codes, often called hydrocodes. Ideally both are done, with many calculations preceding a few experiments to validate code results.

Many other situations can be cited such as bird strikes on canopies and fan blades, design of protective armor systems for personnel and sensitive equipment, erosion of paint from autos struck by small, fast-moving pebbles and assorted road debris as well as applications involving water jets in mining, manufacturing and assorted structural impacts. Like structural dynamics problems, wave propagation situations have a number of common features:

(a) Wave propagation events tend to occur over very short time frames, on the order of nanoseconds to microseconds. This is typically orders of magnitude below the time required for structural response.

(b) Peak strain rates in wave propagation problems tend to be in the range of 10^4–10^6 s^{-1}. After several wave reverberations, typical strain rates will be on the order of 10^2 s^{-1}. Thus, handbook data typically gathered at strain rates of 10^{-4} s^{-1}, are not applicable for such problems. Data for the constitutive models which describe high strain rate behavior must come from wave propagation experiments such as the split-Hopkinson bar, plate impact tests, expanding ring tests, coplanar bar impacts and other sources, as discussed in Chapter 7.

(c) Peak strains will typically be on the order of 60% or more, with some of the more ductile materials approaching strains near 100%. Average strains will be on the order of 20–40%.

(d) Very large pressures are generated when two solids collide. At hypervelocity impact conditions, pressures exceeding material strength by several orders of magnitude are not uncommon. Material strengths are typically in the range of 1–20 kbar (20×10^9 dynes/cm^2). The impact of two steel bodies at 6 km/s will generate a peak pressure of 130 kbar (1.3×10^{12} dynes/cm^2). For problems of long duration, the pressure will eventually decay to a value close to the material strength. For short-duration problems it will dominate the response, with material strength being a second-order effect.

(e) Because of the short duration of the loading and response time, the extent of deformation is *highly localized*. In penetration problems, the nature of the material directly ahead of the striker, as well as its geometry, exerts the greatest influence. Penetration and perforation events are usually completed before the effects of faraway boundary conditions can be felt. Assume, for example, that a striker of arbitrary geometry hits a plate that is 12 cm in diameter and 1 cm thick. Assume that both striker and plate are metallic. If the striking velocity is 1.5 km/s, the particle velocity induced in the plate by the impact will be 0.75 km/s if striker and target are of like materials, somewhat greater if they are not. It will take about 13 μs to penetrate the 1 cm plate. A pressure pulse of magnitude $\rho c v_p$ will propagate from the impact point into the striker and to the radial boundary of the plate. Since the sound speed in most metals is on the order of 5–6 km/s (say 5.5 km/s for the sake of argument), the influence of the plate boundary will not be felt until 44 μs after impact. Hence, material further than about 5–6 striker diameters will play little role in determining response to high-speed impact.

(f) In structural dynamics problems, material failure was an unacceptable level of deformation, failure of a critical member or similar criterion. In wave propagation problems, material failure results in physical separation of materials by a variety of mechanisms. These can include brittle fracture, radial fracture, ductile hole growth, spallation, petalling, plugging and other mechanisms. Unlike fracture mechanics, where one deals with the behavior of a pre-existing crack of known (or postulated) shape, wave propagation problems involve damage where treatment of each crack individually becomes too difficult. According to Seaman [1975], material failure under high-rate loading conditions can be thought of as a four-stage process:

- rapid nucleation of microfractures at a large number of locations in the material;

- growth of the fracture nuclei in a rather symmetric manner;

- coalescence of adjacent microfractures;

- spallation or fragmentation by formation of one or more continuous fracture surfaces through the material.

Even though it is understood that material failure is a micromechanical process coupled to macroscopic response, models which take micromechanical details into account tend to require an inordinate degree of material characterization. Most practical calculations are performed with simpler, continuum failure models of varying degrees of complexity. These will be discussed later.

For background reading in this area consider Meyers [1994], Zukas [1990], Zukas et al. [1982] and Blazynski [1987].

The computer codes (or hydrocodes) which are used for wave propagation problems conform to the relevant physics. Because extremely large distortions are involved, both Eulerian and Lagrangian codes are employed. Indeed, the earliest hydrocodes were finite-difference Euler codes written specifically to solve the problem of hypervelocity impact. Lagrange codes include some sort of technique such as rezoning or dynamic re-definition of colliding surfaces once failed material is removed in order to continue the problem to late times. Without such techniques, Lagrange codes would be limited to the study of the initial stages of an energetic event. Because of the presence of shock waves, only the most basic finite-difference or finite-element schemes (linear, constant strain) are used, as there is no benefit in accuracy but a large penalty in cost to compute discontinuous behavior with high-order polynomials. For the same reason, virtually all hydrocodes use explicit schemes to advance the solution in time. A high-pressure equation of state is incorporated in addition to the constitutive model (which computes deviatoric behavior) to account for the very large pressures encountered.

Failure criteria *in production codes* are based on instantaneous maxima of field variables or account in some way for cumulative damage effects. Micromechanical models are rarely used in production due to the high cost of material characterization. They are generally limited to research applications. Because of the localized nature of the response, boundary conditions tend to be very simple (reflective, transmissive) while initial conditions tend to specify velocity, prescribed deflections and/or surface tractions.

Grove, Rajendran & Walsh [1997] have observed that codes "...have evolved to a point where it is now possible to make predictions without arbitrarily adjusting and manipulating material model parameters." This assumes that the material models are sufficiently accurate to model observed behavior and that the constants for the constitutive model have been determined from wave propagation experiments at strain rates appropriate to the problem. The greatest uncertainty in modeling the material response lies still in determining material failure and post-failure response. Both definitive models and experiments are required in this area. Nevertheless, as Grove, Rajendran and Walsh observe, "...when the failure of a target is dominated by erosion due to very large plastic flow, it is possible to numerically determine the ballistic limits with suitable finite element meshes and robust slide line algorithms."

1.4 Mixed problems.

A good deal of creativity is required for problems that fall in between the above categories, i.e., those which show characteristics of both wave propagation and structural dynamics problems. An example is the impact of a bird against a fan blade of an aircraft engine. The impact energy will be deposited within microseconds after impact. The critical dimension for this part of the problem is the blade thickness. A stress wave will propagate in the thickness direction and be reflected with a change in sign from the rear free surface. If this impact causes catastrophic failure, then this aspect may be treated using wave propagation concepts. If the blade separates into pieces and interacts with other blades or the engine housing, these latter contacts will extend into the structural regime. If the impact does not produce catastrophic failure, then waves in the lateral and transverse directions of the blade will continue to reverberate, eventually inducing deflection in the blade tip. The peak deflection will not be reached until milliseconds after impact. Thus, the response can span several orders of magnitude in time. The initial contact loading must be accurately accounted for, requiring a wave propagation approach by keeping track of individual wave transits. The peak deflection is clearly a structural problem resulting from thousands to hundreds of thousands of wave reverberations. Problems such as these pose a major challenge to existing codes and require considerable insight and experience on the part of the code user.

1.5 Summary.

The primary ingredient for successful simulations of dynamic events is

- *understanding the physical problem* for which a solution is needed;

- *having or obtaining a keen knowledge of the computational tools*;

- *gaining experience in the use of that computational tool*, preferably under the tutelage of a senior engineer or scientist.

Such codes are in no way "black boxes" which can be taken off the shelf and used by anyone. Each has its own characteristics and eccentricities and it takes time—six months to two years—to become thoroughly familiar with a given code.

The best approach to acquiring these skills is to:

(a) Study the basic theory of finite elements, finite differences and other computational techniques used for continuum mechanics, preferably in formal college courses.

(b) Study, either in formal courses or on your own, the basic principles of continuum dynamics, which should include structural dynamics, thermodynamics, material behavior at high strain rates, wave propagation, shock wave physics and mathematical physics.

(c) Acquire fundamental modeling skills under the guidance of experienced colleagues or supervisors.

(d) Become familiar with the constitutive descriptions, numerical models, computational approaches and idiosyncrasies of the particular computer programs you will be using.

The above implies that it would be foolhardy to assign a junior engineer to a structural dynamics or wave propagation code and expect him or her to generate, in short order, the types of results seen in the literature. Commercial codes required 15–25 man-years of development. Their effective use requires a minimum commitment of six months to two years just to set up the code on an in-house computer and learn to use it with a reasonable degree of competency. This is *after* the educational requirements listed in (a) and (b) above have been met. More often than not, considerable on-the-job learning in physics goes on while the code is being mastered. During this period, adherence to a "one person–one code" philosophy is necessary. Equally necessary is contact with other experienced code users or code developers during the learning period. Mandatory is the acquisition of material data for the constitutive equations employed in the code determined from wave propagation experiments at strain rates appropriate for the problems being addressed.

Assuming all this has been done—it almost never is—there remains the problem of rendering the results of code computations in usable graphical form. While most current wave codes are readily transportable, post-processing routines are still highly device-dependent. The color slides and movies produced readily at one installation may require a major investment in graphics hardware at another.

All this is not meant to discourage you from the use of structural dynamics or wave propagation codes. In a later chapter, you will see some outstanding results obtained with present-day codes on technologically difficult problems. It is intended to remind you that the use of codes is a non-trivial exercise. Successful implementation of a computational capability in any organization implies a commitment to manpower, training and investments in hardware and software, which may not be obvious at first glance. Successful use of the codes inevitably entails tradeoffs between accuracy and economy. The rest of the book strives to give you the background necessary to understand the nature of the computer programs used to solve problems in dynamics and to evaluate and use those results intelligently.

REFERENCES

(1996). IRIS Universe, #37, Fall, 1996, Silicon Graphics Corp.

(1987). Proc. 1986 Symp. on Hypervelocity Impact, San Antonio, TX 21–24 Oct. 1986, *Int. J. Impact Eng.*, 5(1–4).

(1990). Proc. 1989 Symp. on Hypervelocity Impact, San Antonio, TX, 12–14 Dec. 1989, *Int. J. Impact Eng.*, 10(1–4).

(1993). Proc. 1992 Hypervelocity Impact Symposium, Austin, TX, 17–19 Nov. 1992, *Int. J. Impact Eng.*, 14(1–4).

(1995). Proc. 1994 Hypervelocity Impact Symposium, Parts I and II, Santa Fe, NM, 17–19 Oct. 1994, *Int. J. Impact Eng.*, 17(1–6).

(1997). Hypervelocity Impact: Proc. 1996 Symposium, Parts I and II, Freiburg, GE, 8–10 Oct. 1996. *Int. J. Impact Eng.*, 20(1–10).

J.D. Achenbach. (1975). *Wave Propagation in Elastic Solids*. New York: Elsevier.

W.F. Adler (ed). (1979). *Erosion: Prevention and Useful Applications*, ASTM-TP-664. Philadelphia: American Society for Testing Materials.
S. Ashley. (1996). Designing safer flywheels. *Mech. Eng.*, November.
M.E. Backman. (1976). *Terminal Ballistics*. Naval Weapons Center, NWC TP 578.
M.E. Backman and W. Goldsmith. (1978). The mechanics of penetration of projectiles into targets. *Int. J. Eng. Sci.*, 16, 1-99.
W.E. Baker. (1973). *Explosions in Air*. Austin, TX: U. of Texas Press.
W.E. Baker, et al. (1983). *Explosion Hazards and Evaluation*. Amsterdam: Elsevier Science Publishers.
S.S. Batsanov. (1993). *Effects of Explosions on Materials*. New York: Springer-Verlag.
A. Bedford and D.S. Drumheller. (1993). *Introduction to Elastic Wave Propagation*. New York: Wiley.
T.Z. Blazynski (ed). (1983). *Explosive Welding, Forming and Compaction*. London: Applied Science Publishers.
T.Z. Blazynski (ed). (1987). *Materials at High Strain Rates*. London: Elsevier Applied Science Publishers.
C.A. Brebbia and V. Sanchez-Galvez (eds). (1994). *Shock and Impact on Structures*. Southampton: Computational Mechanics Publications.
J.H. Brunton and M.C. Rochester. (1979). In C.M. Preece, ed, *Treatise on Materials Science and Technology*, vol. 16, Erosion, New York: Academic Press.
P.S. Bulson (ed). (1994). *Structures Under Shock and Impact III*. Southampton: Computational Mechanics Publications.
J. Carleone (ed). (1993). *Tactical Missile Warheads*. Washington, DC: AIAA.
R. Cheret. (1993). *Detonation of Condensed Explosives*. New York: Springer-Verlag.
P.W. Cooper. (1996). *Explosives Engineering*. New York: Wiley-VCH.
P.W. Cooper and S.R. Kurowski. (1997). *Introduction to the Technology of Explosives*. New York: Wiley-VCH.
B. Crossland. (1982). *Explosive Welding of Metals and Its Application*. Oxford: Clarendon.
A.A. Deribas. (1972). *The Physics of Explosive Hardening and Welding*. Novosibirsk: Nauka.
D.S. Drumheller. (1998). *Introduction to Wave Propagation in Nonlinear Fluids and Solids*. Cambridge: Cambridge U. Press.
W. Fickett and W.C. Davis. (1979). *Detonation*. Berkeley: U. of Calfornia Press.
T.L. Geers and Y.S. Shin (eds). (1991). *Dynamic Response of Structures to High-Energy Excitations*, AMD-vol. 127/PVP-vol. 225. New York: ASME.
S. Glasstone and P.J. Dolan. (1977). *The Effects of Nuclear Weapons*. U.S. Government Printing Office.
D.J. Grove, A.M. Rajendran and K.P. Walsh. (1997). Numerical simulations of tungsten projectiles penetrating titanium targets. In A.S. Khan, ed, *Proc. Intl Conf. on Plasticity*, Juneau, Alaska.

W. Johnson. (1972). *Impact Strength of Materials*. London: Edward Arnold.

W. Johnson and A.G. Mamalis. (1978). *Crashworthiness of Vehicles*. London: Mechanical Engineering Publications, Ltd.

N. Jones. (1989). *Structural Impact*. Cambridge: Cambridge University Press.

N. Jones, C.A. Brebbia and A.J. Watson (eds). (1996). *Structures Under Shock and Impact IV*. Southampton: Computational Mechanics Publications.

S.W. Key. (1985). A comparison of recent results from HONDO III with the JSME nuclear shipping cask benchmark calculations. *Nucl. Eng. Des.*, 85, 15-23.

G.F. Kinney and K.J. Graham. (1985). *Explosive Shocks in Air*. Berlin: Springer-Verlag.

H.S. Levine, J.A. Zukas, D.M. Jerome and Y.S. Shin. (1998). *Structures Under Extreme Loading Conditions—1998*, PVP-vol. 361. New York: ASME.

U.S. Lindholm. (1971). *Tech. Metals Res.*, 5(1).

M.A. Macaulay. (1987). *Introduction to Impact Engineering*. New York: Chapman & Hall.

C.L. Mader. (1979). *Numerical Modeling of Detonation*. Berkeley: U. of California Press.

G. Morandin and E. Nadeau. (1996). Accident impact of a spent fuel dry storage package: analytical/experimental comparison. In Y.S. Shin and J.A. Zukas, eds, *Structures Under Extreme Loading Conditions–1996*, PVP-vol. 325. New York: ASME.

C.L. Mader. (1997). *Numerical Modeling of Explosives and Propellants*. Boca Raton: CRC Press.

I.G. Main. (1993). *Vibrations and Waves in Physics*, 3rd ed. Cambridge: Cambridge University Press.

H.J. Melosh. (1989). *Impact Cratering*. New York: Oxford University Press.

M.A. Meyers. (1994). *Dynamic Behavior of Materials*. New York: Wiley.

L.E. Murr (ed.) (1990). *Shock Waves for Industrial Applications*, Noyes.

T. Nicholas and A.M. Rajendran. (1990). Material characterization at high strain rates. In J.A. Zukas, ed, *High Velocity Impact Dynamics*, Chapter 3. New York: Wiley.

P-A. Person, R. Holmberg and J. Lee. (1994). *Rock Blasting and Explosives Engineering*, Boca Raton: CRC Press.

W. Pilkey and B. Pilkey. (1995). *Shock and Vibration Computer Programs. Reviews and Summaries*. Arlington, VA: The Shock and Vibration Information Analysis Center, Booz, Allen & Hamilton.

C.M. Preece (ed). (1979). *Treatise on Materials Science and Technology, Erosion*, vol. 16. New York: Academic Press.

E.F. Quigley. (1989). EPIC-2 calculated impact loading history for finite element analysis of ballistic shock, Ballistic Research Laboratory, BRL-TR-3058.

E.F. Quigley. (1991). Finite element analysis of non-perforating ballistic impacts using hydrocode-generated loading histories, Ballistic Research Laboratory, BRL-TR-3210.

Y.D.S. Rajapakse and J.R. Vinson. (1995). *High Strain Rate Effects on Polymer, Metal, and Ceramic Matrix Composites and Other Advanced Materials*, AD-vol. 48. New York: ASME.

J.S. Rinehart and J. Pearson. (1963). *Explosive Working of Metals*. New York: McMillan.

L. Seaman. (1975). Fracture and fragmentation under shock loading. In W. Pilkey and B. Pilkey, eds, *Shock and Vibration Computer Programs*. Washington, DC: Shock and Vibration Information Center.

Y.S. Shin (ed). (1995). *Structures Under Extreme Loading Conditions*, PVP-vol. 299. New York: ASME.

Y.S. Shin and J.A. Zukas (eds). (1996). *Structures Under Extreme Loading Conditions—1996*, PVP-vol. 325. New York: ASME.

Y.S. Shin, J.A. Zukas, H.S. Levine and D.M. Jerome. (1997), *Structures Under Extreme Loading Conditions—1997*, PVP-vol. 351. New York: ASME.

G.S. Springer. (1976). *Erosion by Liquid Impact*. New York: Wiley.

C.R. Steele, J. Tolomeo and D. Zetes. (1995). In W. Pilkey and B. Pilkey, eds, *Shock and Vibration Computer Programs. Reviews and Summaries*. Arlington, VA: The Shock and Vibration Information Analysis Center, Booz, Allen & Hamilton.

W.P. Walters. (1990). Fundamentals of shaped charges. In J.A. Zukas, ed, *High Velocity Impact Dynamics*, Chapter 11. New York: Wiley.

W.P. Walters and J.A. Zukas. (1989). *Fundamentals of Shaped Charges*, (reprinted 1997 by CMCPress, Baltimore, MD). New York: Wiley.

R.J. Wasley. (1973). *Stress Wave Propagation in Solids: An Introduction*. New York: Marcel Dekker.

J.C. Watkins and R.A. Berry. (1979). A state-of-the-art literature review of water hammer, EG&G Idaho, Inc., Idaho National Engineering Lab., RE-A-79-044.

J.S. Wilbeck. (1978). Impact behavior of low-strength projectiles, Air Force Materials Laboratory, AFML-TR-77-134.

J.A. Zukas (ed). (1990). *High Velocity Impact Dynamics*. New York: Wiley.

J.A. Zukas and W.P. Walters. (1997). *Explosive Effects and Applications*. New York: Springer-Verlag.

J.A. Zukas, T. Nicholas, H.F. Swift, L.B. Greszczuk and D.R. Curran. (1982). *Impact Dynamics*. New York: Wiley.

J.A. Zukas, J.R. Furlong and S.B. Segletes. (1992). Hydrocodes support visualization of shock-wave phenomena. *Comput. Phys.*, March/April.

FURTHER READING

In addition to the books cited above, there are additional texts which will help broaden your knowledge of high-rate loading phenomena. Some of these are:.

(1980). *Materials Response to Ultra-High Loading Rates*, National Materials Advisory Board, NMAB356, Washington, DC.

W.J. Ammann, W.K. Liu, J.A. Studer and T. Zimmermann (eds). (1988). *Impact: Effects of Fast, Transient Loading*. Rotterdam: A.A. Balkema.

J.R. Asay and M. Shahinpoor (eds). (1993). *High Pressure Shock Compression of Solids*. New York: Springer-Verlag.

A.V. Bushman, G.I. Kanel', A.L. Ni and V.E. Fortov. (1993). *Intense Dynamic Loading of Condensed Matter*. Washington, DC: Taylor & Francis.

P. Cardirola and H. Knoepfel. (1971) and *Physics of High Energy Density*. New York: Academic Press.

P.C. Chou and A.K. Hopkins (eds). (1972). *Dynamic Response of Materials to Intense Impulsive Loading*, Air Force Materials Laboratory.

L. Davison, D.E. Grady and M. Shahinpoor (eds). (1996). *High Pressure Shock Compression of Solids II*. New York: Springer-Verlag.

J. Donea (ed). (1978). *Advanced Structural Dynamics*. London: Applied Science Publishers.

A.G. Evans. (1979). Impact damage mechanics: solid projectiles. In C.M. Preece, ed, *Treatise on Materials Science and Technology, Erosion*, vol. 16. New York: Academic Press.

R.A. Graham. (1992). *Solids Under High-Pressure Shock Compression*. New York: Springer-Verlag.

J. Harding. (1987). The effect of high strain rate on material properties. In T.Z. Blazynski, ed, *Materials at High Strain Rates*. London: Elsevier Applied Science.

W. Johnson and S.R. Reid. (1978). Metallic energy dissipating systems. *Appl. Mech. Rev.*, 31(3).

N. Jones and T. Wierzbicki (eds). (1983). *Structural Crashworthiness*. London: Butterworths.

R.F. Kulak and L. Schwer (eds). (1991). *Computational Aspects of Contact, Impact and Penetration*. Lausanne: ElmePress Intl.

S. Nemat-Nasser, R.J. Asaro and G.A. Hegemier (eds). (1984). *Theoretical Foundation for Large-Scale Computations of Nonlinear Material Behavior*. Dordrecht: Martinus Nijhoff Publishers.

M.L. Wilkins. (1984). Modeling the behavior of materials. In J. Morton, ed, *Structural Impact and Crashworthiness, Conference Papers*, vol. 2. London: Elsevier Applied Science Publishers.

M.L. Wilkins. (1999). *Computer Simulation of Dynamic Phenomena*. Berlin: Springer Verlag.

T.W. Wright. (1983). A survey of penetration mechanics for long rods. In J. Chandra and J.E. Flaherty, eds, *Computational Aspects of Penetration Mechanics*. Heidelberg: SpringerVerlag.

J.A. Zukas. (1987). Stress waves and fracture. In T.Z. Blazynski, ed, *Materials at High Strain Rates*. London: Elsevier Applied Science.

Chapter 2
Wave Propagation and Impact

2.1 Introduction.

Most of our training and experience is geared towards *static* situations, that is, events where things change so slowly that we can think of them as being independent of time. We are aware of dynamic situations of course — automobile crashes, aircraft accidents, industrial accidents, natural phenomena such as tornadoes bearing an assortment of debris, damage from hurricanes and tropical storms to name but a few. Still, to paraphrase Winston Churchill, we manage to experience these and go on as though they had no effect on our approach to solving problems. When dealing with short-term effects, however, we must change our way of looking at the world. Much of the intuition we have developed based on a static view of events fails us when considering truly dynamic problems. It now becomes necessary to be cognizant of two very important factors:

(a) the *rate* at which our observed phenomenon changes;

(b) the fact that information is propagated at a finite speed.

In mechanical systems, this means taking into account both strain rate and wave propagation effects.

What we observe as rigid body motion is actually the net result of many, many wave reflections. Think of it as the long-term solution for problems of intense impact or impulsive loading. Rigid-body impacts, especially as they apply to robotics, are covered very lucidly by Brach [1991]. A very valuable addition to the literature is the text by Stronge [2000]. This approach is useful for a variety of low-speed collisions — hammers on nails, hail on roof tops, athletic balls on floors and walls, screen doors, vehicles, typewriter carriages, dot matrix printers, shot peening, solid particles in multi-phase flows and others. The classic theory follows directly from Newton's laws of motion. More often than not, the problem is formulated in terms of nonlinear differential equations. However, the equations of impulse and momentum are algebraic and almost always linear. This formal approach is coupled to reality through coefficients of restitution and friction. This removes much mathematical unpleasantness while making the theory a tractable and useful engineering tool. Rigid-body mechanics does not take into account the internal energy due to deformation. This is a very significant factor for problems involving intense, short-duration loading. For these problems, we need to look at individual wave transits.

Wave propagation in a solid is governed by both its geometry and its constitution (mechanical properties). Traditionally, two distinct geometries have been used to study wave propagation effects — rods and plates. Both have an apparent simplicity, which disappears once trying anything other than one-dimensional (1D) solutions. Waves in rods and rod-like structures (Fig. 2.1) have been considered to create a state of *uniaxial stress*. The key parameter is the stress along the axis of the rod. The rod supports elastic longitudinal, shear and torsional waves. Once the elastic limit is reached, both elastic and plastic waves propagate. In this configuration, however, it is impossible to reach very high stress states. As strain rate increases, 2D and 3D effects begin to dominate the rod deformation. Our experiments are still valid but our methods of analysis fail us once we lose the uniaxial stress condition. Plasticity and material failure govern the magnitude of the stress that a rod geometry can carry. Typical idealized stress–strain curves used in computations for materials in such configurations are shown in Fig. 2.2. These are derived from typical uniaxial stress experiments routinely performed under quasi-static loading conditions (Fig. 2.3).

In order to examine other states of materials we need to achieve higher levels of stress. For this we change to a plate configuration. Figure 2.4 shows a thin plate, known as the *flyer*, striking a somewhat larger one. Waves will radiate into the stationary plate and the flyer plate in the thickness direction. There will also be waves in the transverse direction. However, until these reflect from the lateral boundary and return to the center, a state of *uniaxial strain* (but 3D stress) will exist there. It will be shown later that, due to

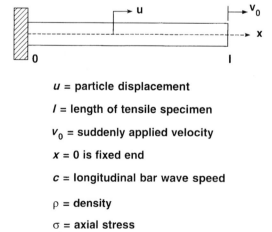

u = particle displacement

l = length of tensile specimen

v_0 = suddenly applied velocity

$x = 0$ is fixed end

c = longitudinal bar wave speed

ρ = density

σ = axial stress

Figure 2.1 Rod geometry for wave propagation studies.

Wave Propagation and Impact 35

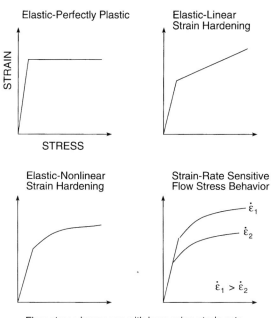

Figure 2.2 Idealizations of stress–strain behavior for rod geometries.

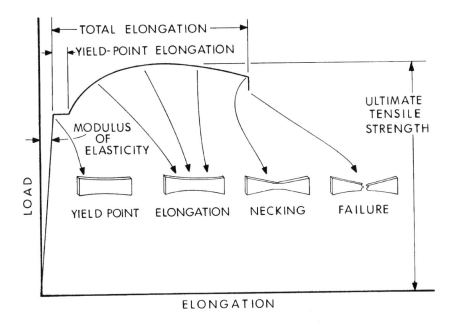

Figure 2.3 Quasi-static stress–strain curve.

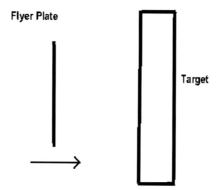

Figure 2.4 Plate impact geometry.

this change in geometry, we can now achieve hydrodynamic stress states (or pressures) orders of magnitude higher than the strength of the material at strain rates approaching 10^5 s^{-1}, as compared to maximum strain rates in rod configurations of between 10^2 and 10^3 s^{-1}. We will explore the potential of plate geometries in Chapter 3. For the remainder of this chapter, we focus on stress states in rods. For additional information on this topic see Kolsky [1963], Towne [1967], Graff [1975], Main [1993], Zukas et al [1982], Zukas [1990], Bedford and Drumheller [1994] and Drumheller [1998].

2.2 Wave propagation in rods and plates.

Let us introduce the notion of wave propagation by considering a simple yet commonplace problem.

A semi-infinite body is subjected to impact by a slender, cylindrical striker, long enough so that wave reflections from its rear surface need not concern us. For the moment, assume that striker and target are made of the same material.

a) *Describe the wave motion set up in the body.*

The situation we will discuss is depicted in Fig. 2.5. Different types of elastic waves can propagate in solids, depending on how the motion of the particles in a solid is related to the direction of propagation and on the boundary conditions. Note that when we refer to "particle" here we are not talking about the motion of atoms. Elasticity theory is based on the assumption of a continuum. The effects of individual atomic motions are seen only in the aggregate, each material particle being composed of a sufficiently large number of

Wave Propagation and Impact 37

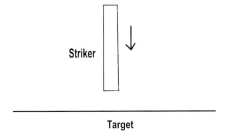

Figure 2.5 Striker impacting semi-infinite surface.

atoms so that it can be viewed as a continuous entity. The most common types of elastic waves are:

Longitudinal waves: In such waves the particle motion is back and forth along the direction of propagation. The particle velocity v_p is parallel to the wave velocity c. If the wave is compressive, they have the same sense. If the wave is tensile, they have opposite senses. In the literature, longitudinal waves are also referred to as irrotational waves, push, primary or P waves. In infinite and semi-infinite media, they are known as dilatational waves (Fig. 2.6).

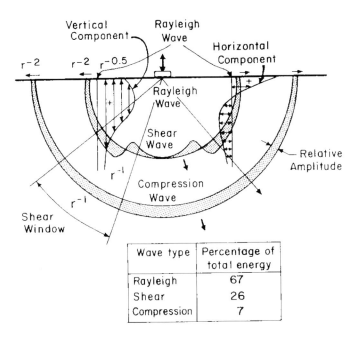

Figure 2.6 Propagation of compressive, shear and surface waves in semi-infinite media [Meyers, 1994]. This material is used by permission of John Wiley & Sons, Inc.

Distortional (or shear) waves: Here, the motion of the particles conveying the wave are perpendicular to the direction of propagation of the wave, Fig. 2.5. There is no resulting change in density and all longitudinal strains $\varepsilon_{11}, \varepsilon_{22}, \varepsilon_{33}$ are zero.

Surface waves: Surface waves are analogous to waves on the surface of water. Objects floating on the water act as markers for the particles of water and move both up and down and back and forth, tracing out elliptical paths as the water moves by. This type of wave occurs only in a narrow region close to the surface and the particle velocity v_p decays exponentially away from the surface. In solids, surface waves are called Rayleigh waves. They are a particular case of interfacial waves when one of the materials has negligible density and elastic wave velocity. These are also known as transverse or head waves (Fig. 2.6).

Interfacial (Stoneleigh) waves: When two semi-infinite media with different properties are in contact, these waves form at their interface.

Waves in layered media (Love waves): These are of particular importance in seismology. Earthquakes can produce waves in which the horizontal component of displacement can be significantly larger than the vertical component, a behavior not consistent with Rayleigh waves. The earth is composed of layers with different properties and special wave patterns emerge.

Bending (or flexural) waves: These waves involve propagation of flexure in one or two-dimensions in structural members such as beams, plates and shells.

In isotropic elastic solids in which body forces (gravitation and moments) are absent, longitudinal and shear waves are the most significant.

Now to the specifics of the question. It might be best to start by asking what would happen if impact occurred between two collinear rods, each moving towards the other with the same velocity. Figure 2.7 shows results of a *ZeuS* finite element wave propagation code calculation for this situation at 0.114 μs after impact. Note that a planar (1D) compressive wave propagates along the length of each rod, emanating from the interface. Because the rods are made of the same material (aluminum in this calculation), the intensity of the propagating stress pulse will be $\sigma = \rho c v_p$. However, these rods are 2D bodies. The two-dimensionality comes into play through the boundary conditions. The lateral surfaces of the rod are free surfaces. Free surfaces, by definition, carry no stress. The elementary (1D) theory of wave propagation in elastic bodies ignores these effects. The theory serves us well in many cases and thus we use it frequently. Keep in mind, though, that the 2D response represents reality. In special cases such as short rods, the response is wholly 2D and the elementary theory no longer applies. Thus, a compressive pulse acting at the lateral boundary must be reflected as a tensile pulse in order to satisfy the stress-free boundary condition. Therefore, unlike the elementary theory, we will have a compressive pulse moving up the rod but behind it there will exist a complex 2D stress state consisting of the interaction of tensile and compressive stress waves oscillating

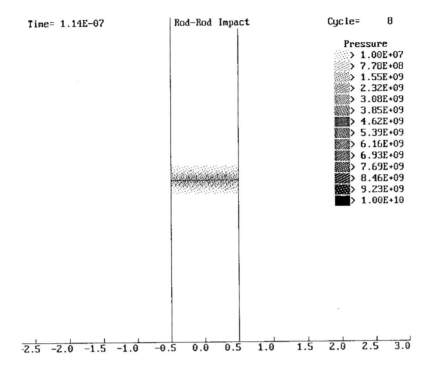

Figure 2.7 Planar waves in colliding rods just after impact.

radially. You can see this procedure starting in Fig. 2.8 at 0.7 μs after impact and pretty well developed in Fig. 2.9 at 6 μs after impact where, behind the compression front, there are regions of tension and compression. The physics is quite clear, but keeping track of all the interacting waves required a book-keeping tool like a computer.

What happens now if we expand the radial and axial boundaries of the lower rod so that it becomes, in effect, a semi-infinite solid. Right after impact, the same conditions will exist in rod and target as depicted in Fig. 2.7, namely a planar wave will propagate in both bodies. It is difficult and expensive to model numerically semi-infinite media with finite elements unless you are using boundary elements. In practice one sets up calculations where the radial boundary is sufficiently far removed that any signal returning from that boundary comes long after we have numerically interrogated the body for its response to a stress pulse or pulses. In effect, then, we have modeled the behavior of a semi-infinite body using large but finite geometry. All this is to reduce the cost of a calculation. Even greater savings can be obtained by imposing appropriate boundary conditions, using boundary elements, etc. A portion of the grid is shown in Fig. 2.10, while Fig. 2.11 shows pressures shortly after impact. You can still see the planar wave in the rod. In the target, though, you have a free surface where contact with the rod ceases

40 Introduction to Hydrocodes

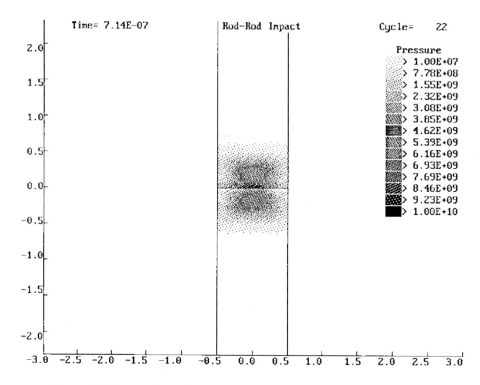

Figure 2.8 Progression of compressive wave in rods with release waves from lateral surfaces.

($r > D/2$). Again, relief (tensile) pulses must emanate from this region. Thus, you see the edges of the planar pulse being nibbled away as it were (tensile pulses reducing the intensity of the compressive pulse) and in short order (Fig. 2.12) we have a spherical compressive pulse moving in the target. Because longitudinal waves move with a greater particle velocity than shear waves, compressive effects will dominate at early times. By 2 μs after impact sufficient time has elapsed to feel the effects of tensile relief waves from the target free surface and shear waves as well, although in plots such as these the effects of shearing distortions are not evident (Fig. 2.13). By 5 μs, our compressive pulse has reached a radial boundary. Were it not there, the pulse would continue to expand, dissipating its energy through heating of the target material and modest free surface displacements (Fig. 2.14). Note that only compressive pressures are depicted in the plot. The white regions thus represent portions of the target in tension.

Despite the fact that the pressure generated by the impact has spread over a considerable portion of the target, the deformations, which ensue are highly localized. This is shown in Fig. 2.15, which plots the distribution of *effective plastic strain* in target and striker. Note that the regions of high plastic deformation, up to 140% strain, are

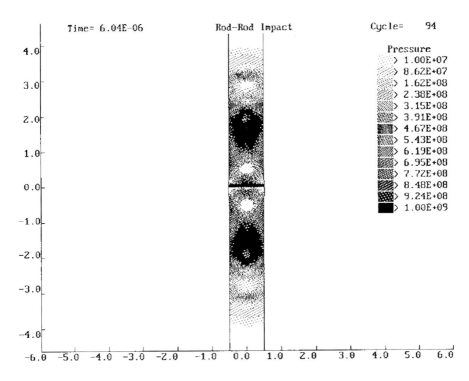

Figure 2.9 Existence of complex 2D stress state.

localized within 1–2 projectile diameters in both bodies. As a rule of thumb, the most severe deformations will occur within 3 characteristic lengths (in this case, projectile diameters) of the impact location, going up to perhaps 5–6 for hypervelocity impacts ($v_p/c > 1$, or about 6 km/s for most metals).

b) *Describe the wave motion set up in the striker.*

In effect, we have already done this in the above discussion. However, those few paragraphs contained a lot of information by implication. Let us look at it again from another perspective.

Consider Fig. 2.16. It depicts a bar, which has been subjected to a disturbance at one end. This could be a suddenly applied pressure pulse, an impact by another bar, an impact against a rigid barrier, and so on. Note that two velocities are shown, c and v. The first, c, is called the velocity of propagation of the disturbance, or, more often, the sound speed. It is often referred to as a material property, but its numerical value depends on geometry and boundary conditions, as will shortly be shown. The distance a disturbance propagated

42 *Introduction to Hydrocodes*

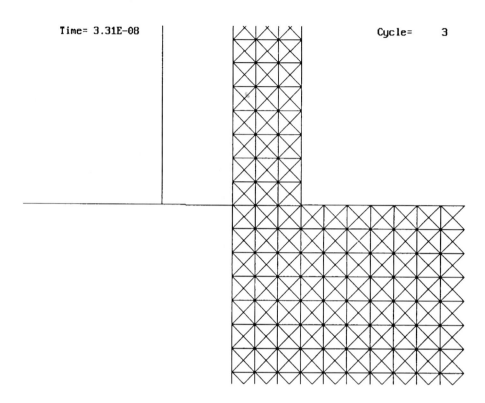

Figure 2.10 Finite-element grid for rod impact calculation using the *ZeuS* code.

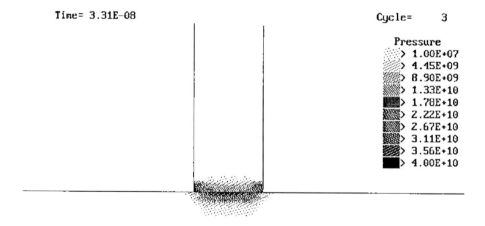

Figure 2.11 Planar waves propagating in rod and target. Note the effect of relief waves from the target free surface.

Wave Propagation and Impact 43

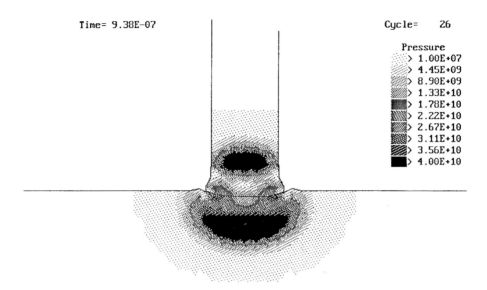

Figure 2.12 Stress wave propagation in rod and target.

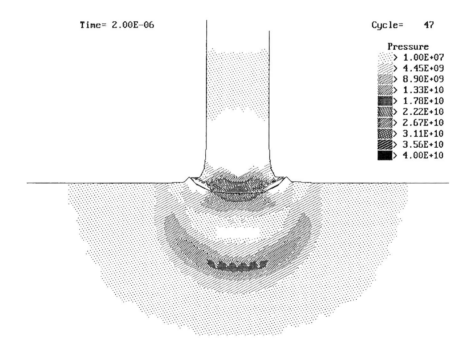

Figure 2.13 Wave propagation showing relief wave buildup behind compressive front.

44 *Introduction to Hydrocodes*

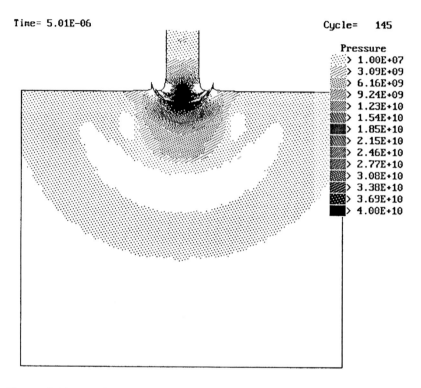

Figure 2.14 Development of tensile region behind compressive wave due to boundary effects.

in a time interval Δt is $c\Delta t$. All material particles engulfed by the disturbance out to the distance $c\Delta t$ are aware that something has happened. Beyond that distance, the material particles are in their original state and unaware of what is coming. The second velocity depicted, v, is referred to as the particle velocity. The distance $v\Delta t$ is the actual deformation imposed on the material by the disturbance. Why two velocities? Inertia. A particle at rest subjected to an unbalanced force will move, but not instantly. It may need to overcome its own inertia, it may need to wait until neighboring particles move and constraints on its own motion are freed, etc. Consider a rather imperfect yet useful analogy. A long line of cars travels down a one-lane road with a traffic light at an intersection, which is initially green. At some time, the light turns red. Think of this as the equivalent of having a disturbance applied at one end of our rod. This red signal is seen be all cars in line almost at once. Think of this as the disturbance, or stress pulse, in the rod. The car closest to the intersection slams on the brakes. The second car brakes as well but because of the sudden stop of the first vehicle, crashes into it. The third car likewise crashes into the second and so on. In effect, another wave is now propagating down

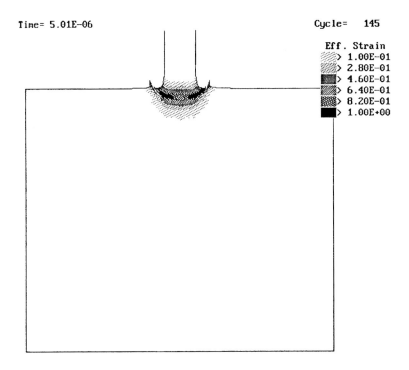

Figure 2.15 Distribution of plastic strain in rod impact.

the line of cars, the wave of deformed cars but at a speed much slower than the observed speed of the traffic light. Think of the red light as covering a zone of length $c\Delta t$ while the zone of moving cars constitutes the distance $v\Delta t$.

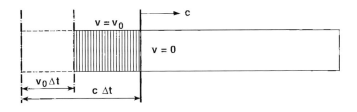

At t = 0, a compressive pulse strikes bar at velocity

At t = Δt
- Compressive wave has moved distance c Δt
- Length of bar set in motion is v_0 Δt

Figure 2.16 Depiction of sound velocity and particle velocity in an impacted bar.

46 *Introduction to Hydrocodes*

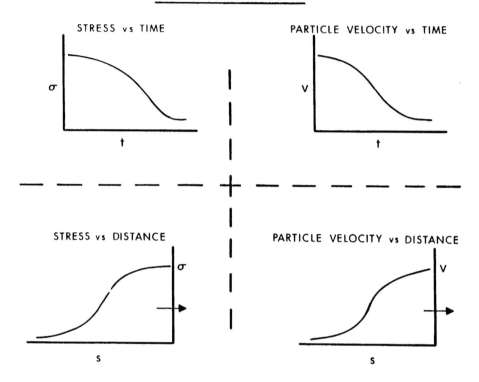

Figure 2.17 Various representations of disturbances moving in impacted bodies.

These pulses can be depicted in a number of ways (Fig. 2.17). The choice depends of which aspect of a problem we are trying to solve. For convenience we can plot:

(a) stress vs. time;

(b) particle velocity vs. time;

(c) stress vs. distance;

(d) particle velocity vs. distance.

Referring to Fig. 2.16, let us determine the stress at the interface upon impact. We will do this using the most elementary 1D theory of elastic wave propagation. This entails several assumptions:

Wave Propagation and Impact

(a) We assume that we are dealing with a slender bar, one of circular cross-section whose length is at least 10 times greater than its diameter.

(b) In addition, we neglect the effects of

transverse strain;

lateral inertia;

body forces;

internal friction.

The relationship between the longitudinal stress at any point in the bar and the longitudinal particle velocity v comes from Newton's second law

$$F\Delta t = mv, \tag{2.1}$$

where F is the longitudinal force acting on a given cross-section; Δt, the time the force acts; m, the mass it acts against and v, the velocity imparted to m by F. Since

$$\sigma = \frac{F}{A}, \tag{2.2}$$

and

$$m = \rho A dl, \tag{2.3}$$

where dl is the distance the pulse has moved in time dt, we can write

$$\sigma A \Delta t = \rho A dl dv, \tag{2.4}$$

or

$$\sigma = \rho \frac{dl}{dt} dv, \tag{2.5}$$

but dl/dt is just the speed of the pulse, c. Therefore, we can write the stress intensity at the interface as

$$\sigma = \rho c (\Delta v_p). \tag{2.6}$$

Remember — the equation uses the *change* in *particle velocity*. If the initial particle velocity was zero, then it is just good old $\sigma = \rho c v_p$. However, if one of the bodies involved in a collision is already moving, you need the change in particle velocity.

These results are based on elementary 1D wave theory. How do they differ from experiments and more complicated analyses? That can be answered with the help of Figs. 2.18 and 2.19. The elementary solution for wave propagation in rods is a perfectly square pulse, the dashed line in Fig. 2.18. In reality, the response of the rod is two-dimensional. The kinetic energy deforming the rod will also cause deformation in the radial direction. Pochhammer [1876] and Chree [1886; 1889a,b] were among the first to calculate wave speeds in cylindrical rods. The differences between the elementary solution and experiments or 2D analyses are

(a) a finite rise time in the stress pulse;

(b) oscillations, or ringing (sometimes called Pochhammer–Chree ringing), behind the leading edge of the pulse. Skalak [1957] was able to account for transverse strain effects and derive a tractable analytical solution which closely matches experimental observations.

Figure 2.19 shows the development of the stress pulse in a long elastic bar with length-to-diameter (L/D) ratio of 50 striking a rigid barrier at 6.1 m/s. This is a 2D

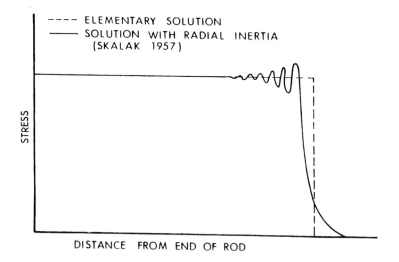

Figure 2.18 Comparison of exact and elementary solutions for semi-infinite bar.

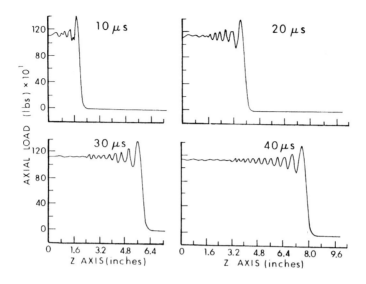

Figure 2.19 EPIC-2 results at various times for $L/D = 50$ bar impacting a rigid barrier.

numerical solution for a bar of circular cross-section ($r = 0.254$ cm, $L = 2.54$ cm) showing the development of the stress pulse over time. Note that at long times after impact the elementary solution is obtained far behind the head of the pulse. The elementary solution is then a very useful estimation tool for peak stress and particle velocities achieved on impact, provided that the assumptions of the 1D solution are not violated. When they are, recourse must be made to more complex calculations or to numerical techniques. Figures 2.20–2.22 show 2D numerical calculations for a short steel bar, four diameters long, impacting a rigid surface at 10 m/s [Zukas, Nicholas, Swift, Greszczuk & Curran, 1982]. Plotted is the axial force in the bar, normalized by $\rho c v$, against position in the bar, normalized by its length. The deviations from the elementary solution (dashed line) and the significance of radial inertia are readily apparent, especially once reflected waves from the read of the bar begin to influence the solution (Fig. 2.22).

2.3 Flies in the ointment.

Thus far, we have treated the sound speed in the bar as though it were a material property, to wit, a constant. In reality it is a function of geometry and boundary conditions. What is worse, there are several different sound speeds — one for

50 Introduction to Hydrocodes

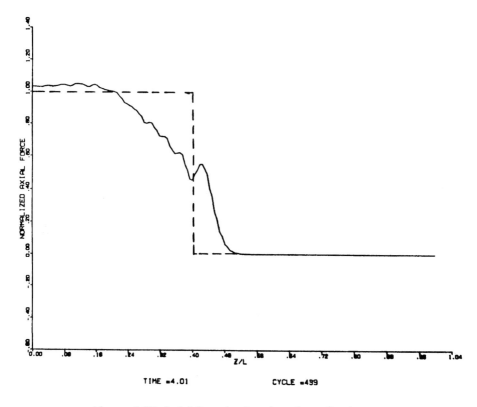

Figure 2.20 Axial force in short bar, 4 μs after impact.

longitudinal waves, one for shear waves, one for bulk solids, one for bars. Let us clear this up before it becomes too complicated.

Figure 2.23(a) shows a section of a round bar through which a longitudinal *stress pulse* travels. Ignore the effects of inertia and friction, assume the bar is long enough and that we are looking at the wave motion at a point far away from the application of the loading.

This avoids messy 3D complications at only slight sacrifice in reality, Fig. 2.23(b). Focus on a section of the bar and apply momentum conservation and continuity. For the sake of simplicity, assume a compressive pulse. Exactly the same thing can be done with a tensile pulse provided one is careful with the sign convention. Here we assume that a compressive stress is positive, as is motion in the positive *x*-direction (to the right). Assuming further that the bar is elastic, we arrive at the well-known 1D wave equation and its solution in the form of traveling waves. The constant c_L is known as the "longitudinal wave speed," often denoted by just plain *c*, and is given by the square root of the bar's elastic modulus divided by its density

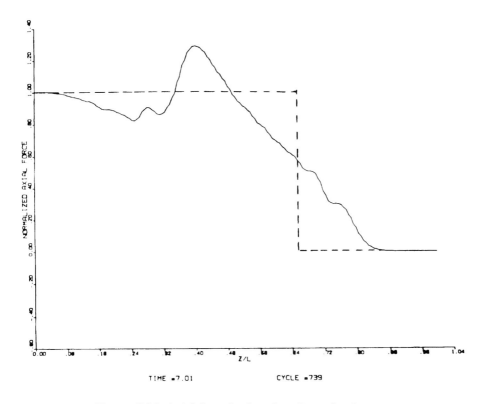

Figure 2.21 Axial force in short bar, 7 μs after impact.

$$c_{\rm L} = \sqrt{\frac{E}{\rho_0}}. \tag{2.7}$$

By an exactly analogous process (Figs. 2.24(a) and (b)) we can consider shear (torsional) waves and determine that their sound speed is given by

$$c_{\rm s} = \sqrt{\frac{G}{\rho_0}}. \tag{2.8}$$

Bars are bounded media. What if we have to deal with large, unbounded media when considering earthquakes or the impact of small strikers onto large surfaces? Do these equations still apply? No. In unbounded media we deal with either two or three dimensions and now can no longer ignore Poisson ratio effects. The analogous equations

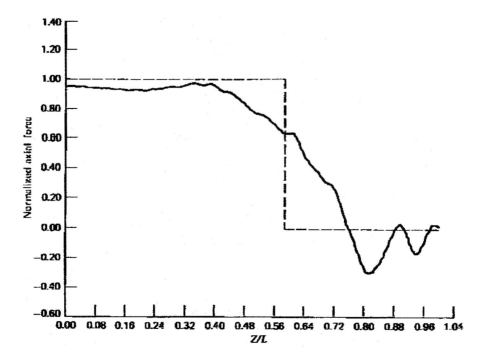

Figure 2.22 Axial force in short bar, 14 μs after impact.

for unbounded media are

$$c_L = \sqrt{\frac{E(1-\nu)}{\rho_0(1+\nu)(1-2\nu)}}, \qquad (2.9)$$

and

$$c_s = \sqrt{\frac{G}{\rho_0}} = \sqrt{\frac{E}{2\rho_0(1+\nu)}}, \qquad (2.10)$$

since, for isotropic materials, $E/G = 2(1+\nu)$.

Professor William Johnson's delightful book [Johnson, 1972] has derivations for the wave equation for a bar constrained to deform under conditions of plane strain.

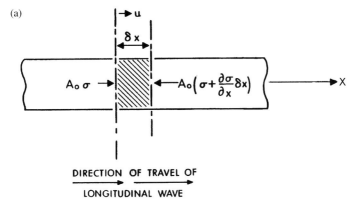

Figure 2.23 (a) Force balance on rod cross-section. (b) Longitudinal sound speed in an elastic bar from elementary theory.

The resultant sound speed is

$$c'_L = \sqrt{\frac{E}{\rho(1-\nu^2)}}. \tag{2.11}$$

The difference between this and the bar velocity is

$$\frac{c'_L}{c_L} = \frac{1}{\sqrt{1-\nu^2}}. \tag{2.12}$$

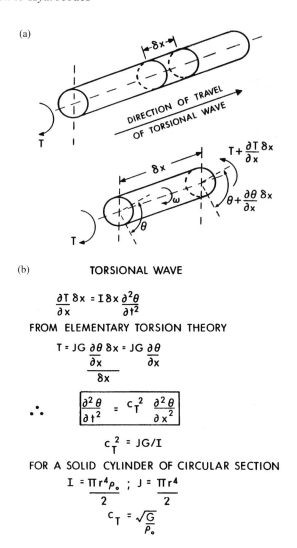

Figure 2.24 (a) Torsional wave propagation. (b) Shear sound speed.

For most metals, Poisson's ratio ν is between 0.25 and 0.33. This means that the above ratio will range from 1.03 to 1.06 for all practical problems.

Johnson further derives the equation for sound speed for a uniform bar constrained to have zero transverse deformation. The expression is

$$c_L'' = \sqrt{\frac{E(1-\nu)}{\rho(1+\nu)(1-2\nu)}}. \qquad (2.13)$$

For Poisson's ratios of 0.25–0.33, the ratio of c_L''/c_L ranges between 1.1 and 1.22. However, the real fun begins if Poisson's ratio reaches 0.5. Then we have

$$c_L = 1; \qquad c_L' = 1.15; \qquad c_L'' = \infty. \tag{2.14}$$

The moral of the tale is this: any so-called "sound speed" is a function of geometry and constraints as well as material characteristics. If you stay away from extreme conditions such as Poisson's ratio approaching 1/2, then for all practical purposes you need to worry about only about the bar sound speed $c_L = \sqrt{E/\rho}$ in long, slender members or the bulk sound speed for extended members such as plates. Most tabulations will indicate which sound speed they list. If they do not, assume it is the bulk sound speed if you are reading shock physics literature, the bar speed if you are looking at the engineering literature.

2.4 Bending waves.

Do bending waves satisfy the wave equation? Look at Fig. 2.25 (a) and (b). The resulting equation of motion for a bar in flexure has a fourth-order derivative in displacement. It is not the hyperbolic wave equation and it does not admit the traveling wave solution. Hence, bending waves are in a class by themselves.

2.5 Wave reflections and interfaces.

Consider the problem of a slender rod impacting a rigid barrier, as shown in Fig. 2.26. The sequence of events is as follows:

(a) After impact, a compressive wave of intensity $\rho c_L v_0$ moves into the bar. At $0 \leq t \leq l/c_L$ the particles engulfed by the wave are brought to rest.

(b) At $t = l/c_L$, the bar is stationary but in compression. All of the kinetic energy has been converted to strain energy.

$$\text{IE} = (\text{volume})\frac{\sigma^2}{2E} = \frac{Al}{2E}(\rho c_L v_0^2) = \frac{1}{2} A l \rho v_0^2. \tag{2.15}$$

$$\text{KE} = \frac{1}{2}(\text{mass})v_0^2 = \frac{1}{2} A l \rho v_0^2, \tag{2.16}$$

56 *Introduction to Hydrocodes*

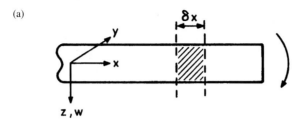

(a)

(b) FLEXURAL WAVES IN BEAMS

EQUATION OF MOTION IN $\bar{z}$-DIRECTION:

$$-\rho_0 A_0 \delta x \frac{\partial^2 w}{\partial t^2} = \frac{\partial Q \delta x}{\partial x}$$

$$Q = EI \frac{\partial^3 w}{\partial x^3}$$

$$\boxed{\frac{\partial^2 w}{\partial t^2} = -c^2 k^2 \frac{\partial^4 w}{\partial x^4}}$$

$c^2 = E/\rho_0$

k = RADIUS OF GYRATION ($I = A_0 k^2$)

NOTE: $w = f(x-ct)$ or $w = f(x+ct)$ is <u>not</u> a solution to this equation.

Figure 2.25 (a) Bending wave. (b) Governing equations for bending waves.

(c) At $t = l/c_L$, the compressive wave is reflected from the rear free surface as a tensile wave. The tensile wave acts as an unloading wave, canceling the effects of the incident compressive wave.

(d) At $t = 2l/c_L$, the bar is completely stress-free. The reflected tensile wave has conferred a speed v_0 on the particles whose direction is opposite to that at impact, therefore, at $t = 2l/c_L$, the bar moves away at equal but opposite speed.

Wave Propagation and Impact 57

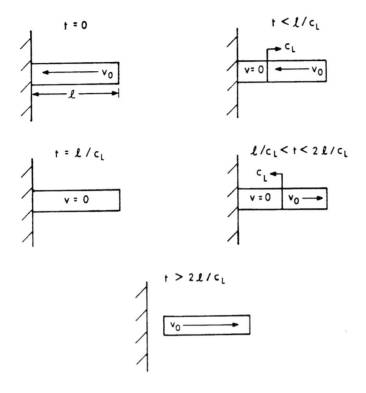

Figure 2.26 Velocity distribution in bar impact.

A formal treatment of this problem may be found in the books by Goldsmith [1960] and Achenbach [1973]. The above technique, however, is useful for visualizing the implementation of the concepts we have been discussing.

Thus far we have considered rods and unbounded media. In practice we encounter finite geometries, sometimes made of multiple materials or irregular shapes. What happens when waves hit free surfaces, material interfaces or geometric discontinuities? Let us consider several possibilities.

a) *Discontinuous cross-sections and different materials.*

Figure 2.27 shows the case of a pulse propagating to the right in a bar with a discontinuous cross-section and made of different materials, for convenience labeled materials 1 and 2. The respective cross-sectional areas of the bar are labeled A_1 and A_2. At the interface of regions 1 and 2 of the bar, the incoming wave of intensity σ_I

58 Introduction to Hydrocodes

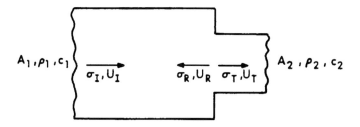

Figure 2.27 Stress waves in bars with discontinuous cross-sections.

and particle velocity v_I will be partially transmitted and partially reflected, as denoted by the subscripts R and T in the figure. Two conditions must be satisfied at the interface:

(a) The forces in both bars at the interface must be equal.

(b) The particle velocities at the interface must be continuous.

Condition (a) gives

$$A_1(\sigma_I + \sigma_R) = A_2 \sigma_T. \tag{2.17}$$

Condition (b) gives

$$v_I - v_R = v_T, \tag{2.18}$$

or, using $\sigma = \rho c v$

$$\frac{\sigma_I}{\rho_1 c_1} - \frac{\sigma_R}{\rho_1 c_1} = \frac{\sigma_T}{\rho_2 c_2}. \tag{2.19}$$

Solving for the transmitted and reflected stresses in terms of the incident stress gives

$$\sigma_T = \frac{2 A_1 \rho_2 c_2}{A_1 \rho_1 c_1 + A_2 \rho_2 c_2} \sigma_I, \tag{2.20}$$

$$\sigma_R = \frac{A_2 \rho_2 c_2 - A_1 \rho_1 c_1}{A_1 \rho_1 c_1 + A_2 \rho_2 c_2} \sigma_I. \tag{2.21}$$

There are several interesting implications coming from these relations:

(a) Let A_2/A_1 approach 0. Alternatively, let $\rho_2 c_2 = 0$. The end of the rod is effectively free. There will be no transmitted wave and the reflected wave $\sigma_R = -\sigma_I$. Thus, *at a free surface*,

compression reflects as tension (and vice versa);

the particle velocity at the free surface doubles.

(b) Let A_2/A_1 approach infinity. The end of the rod is effectively fixed. Then $\sigma_R = \sigma_I$; $\sigma_T = 0$. Thus, *at a fixed surface*,

compression reflects as compression;

particle velocity at the surface is zero.

(c) For no wave reflection to occur from the discontinuity,

$$\sigma_R = 0; \qquad A_2 \rho_2 c_2 = A_1 \rho_1 c_1; \qquad \sigma_T = \sigma_I \sqrt{E_2 \rho_2 / E_1 \rho_1}. \tag{2.22}$$

Let us consider several applications of these equations.

b) *Wave propagation involving discontinuous cross-sections.*

Johnson [1972] has a marvelous example of wave propagation in stepped bars. The situation is shown in Fig. 2.28. Assume a wave traveling from left to right. For the sake of generating some numbers, let $A_1/A_3 = 4$ and $A_1/A_2 = A_2/A_3 = 2$. Take all sections to be made of the same material so that $\rho_1 c_1 = \rho_2 c_2 = \rho_3 c_3 = \rho c$. The transmitted stress in

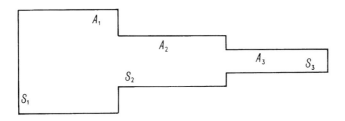

Figure 2.28 Stress magnification in stepped bars [Johnson, 1972].

60 Introduction to Hydrocodes

the section of area A_2 is then given by

$$\frac{\sigma_{T_2}}{\sigma_I} = 2\frac{A_1}{A_1 + A_2} = \frac{2 \times 2}{2 + 1} = \frac{4}{3}. \tag{2.23}$$

The transmitted stress in the section with area A_3 is

$$\sigma_{T_3} = \frac{2A_2}{A_2 + A_3}\sigma_I = \frac{2 \times 4}{4 + 1}\sigma_I = 1.78\sigma_I. \tag{2.24}$$

If the step A_2 is omitted so that the change in area from A_1 to A_3 occurs at once, then

$$\sigma_{T_3} = \frac{2A_1}{A_1 + A_3}\sigma_I = 1.60\sigma_I. \tag{2.25}$$

Inclusion of the extra step results in an increase in stress by about 10%. Thus, for transmission of shock loads, an abrupt change of section is better than a gradual one.

c)   *Layered media.*

Layered media are frequently encountered in practice. Nuclear reactor containment structures are often made of layers of steel and concrete. Storage bunkers and protective shelters may be made of monolithic concrete. However, if the loading is severe, the thickness required to mitigate a blast or impact load may be so great as to make the construction extremely expensive. A combination of concrete with some energy absorbing material may well have the same effect in mitigating stress pulses but at considerably lower construction costs.

In a laboratory environment, tests are often conducted on structures consisting of multi-layered plates, each with its own acoustic impedance (ρc). Depth of penetration tests are also performed on stacks of plates, each of the same material, because the total thickness of the material required may not be available commercially, prohibitively expensive, or both. Not unusual are protective structures consisting of two or more plates with gaps between them. A typical example is a Whipple bumper [see Whipple [1952]; Rajendran & Elfer [1989] and Zukas & Segletes [1991] for details], which is used for protection of space structures. The first plate, a sacrificial bumper, serves to breakup an incoming projectile. An air gap allows the debris to spread so that the main wall now encounters a distributed load instead of a potentially fatal point impact. Figure 2.29 shows the breakup of an incoming projectile by the first plate of the Whipple bumper

Wave Propagation and Impact 61

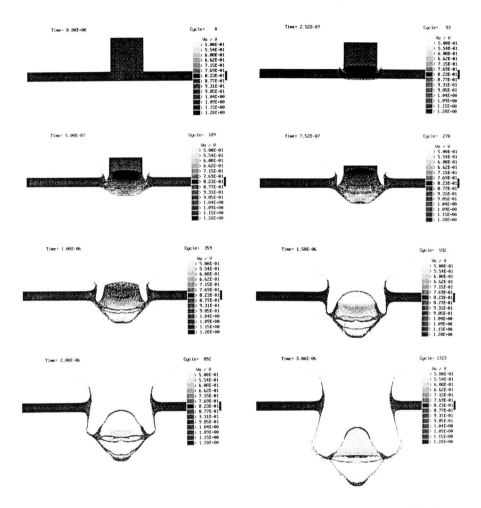

Figure 2.29 *ZeuS* code finite element simulation of bumper perforation and debris cone formation. Dots represent free-flying nodal masses associated with failed elements [Zukas & Segletes, 1991].

[Zukas & Segletes, 1991]. The incoming projectile is reduced to rubble by about 4 μs after impact. The debris then spreads and propagates until it encounters the second plate, which then sees a distributed load. Many bumper designs are in use today, some with up to five plates inserted before the rear wall.

Figure 2.30 shows a situation where a plane wave impacts a three-layered structure with acoustic impedances in the ratio of 1:2:4 [Rinehart, 1960]. The wave is transmitted through medium 1 with intensity σ_I. When it arrives at the interface between medium 1 and medium 2, part of the wave will be transmitted and part reflected. The intensities of

62 Introduction to Hydrocodes

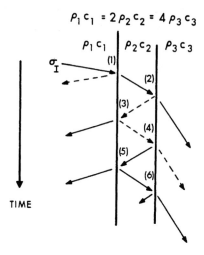

Figure 2.30 Stress wave transmission across laminated plates.

TRANSMISSION ACROSS LAMINATED STRUCTURES

(1) $\sigma R_1 = -1/3 \sigma_I$
$\sigma T_1 = 2/3 \sigma_I$

(2) $\sigma I_2 = \sigma T_1$
$\sigma R_2 = -2/9 \sigma_I$
$\sigma T_2 = 4/9 \sigma_I$

(3) $\sigma I_3 = \sigma R_2$
$\sigma R_3 = -2/27 \sigma_I$
$\sigma T_3 = -8/27 \sigma_I$

(4) $\sigma I_4 = \sigma R_3$
$\sigma R_4 = 2/81 \sigma_I$
$\sigma T_4 = -4/81 \sigma_I$

(5) $\sigma I_5 = \sigma R_4$
$\sigma R_5 = 2/243 \sigma_I$
$\sigma T_5 = 8/243 \sigma_I$

(6) $\sigma I_6 = \sigma R_5$
$\sigma R_6 = 2/729 \sigma_I$
$\sigma T_6 = 4/729 \sigma_I$

Figure 2.31 Intensities of incident and reflected stresses at points 1–5 of Fig. 2.30.

transmitted and reflected stresses can be computed using the simple relations given above (here all cross-sectional areas are identical and drop out of the formulas). This is point (1) of Fig. 2.30. The intensities are given in Fig. 2.31. The transmitted wave now propagates through medium 2 until it reaches the interface between media 2 and 3 [point (2)]. Again, there is transmission and reflection. Because of the values that we have selected for the acoustic impedances of the materials, the wave reflected into media 2 at point (2) will be a tensile wave. Once this wave reaches the interface between media 2 and media 1 [point (3)], transmission and reflection occurs again. The transmitted wave becomes a compression wave, the reflected wave remains a tensile wave (strictly because we chose $\rho_1 c_1 = 2\rho_2 c_2 = 4\rho_3 c_3$ — any other values would change both the nature and the intensity of the waves). At point (4) and subsequent points, the process is repeated until the waves are damped out. Look at Fig. 2.31. Note that at point (1), one-third of the incident wave is reflected while two-thirds is transmitted. By the time that the wave enters medium 3 at point (2), it is at less than half of its original intensity. Thus, layered media can be very useful in attenuating stress pulses. Tedesco & Landis [1989] made an extensive study of wave propagation in layered systems. They found that, if care is taken to choose acoustic impedances which do not give rise to internal spall, layered media can be very effective in reducing transmitted stress levels and can serve to reduce the total thickness of the structure at great savings in material and construction costs.

What if there are gaps between the plates. No momentum can be transferred to the neighboring (or receptor) plate until the gap is closed. Until the gap closes, the transient wave is reflected from the free surface into itself. If there is sufficient motion in the plate to close the gap, then reflection and transmission occur as above.

Segletes & Zukas [1989] considered wave propagation and deformation in multi-layered plates consisting of the same material and compared these to the response of a single layer. Nixdorff [1984] as well as Woodward & Crouch [1988] developed analytical models to study laminated plate behavior, which is frequently encountered in the study of terminal ballistics. Netherwood [1979] investigated several methods for obtaining penetration vs. time data within steel targets for normal impacts at velocities of 1 km/s. Several different approaches were used to try to measure arrival times of the penetrator. These included drilling holes in the target at various locations along the thickness as well as cutting the monolithic plate into several layers, inserting gauges and reassembling the layers. Total target thickness varied between 15–50 mm.

The greater the alterations to the monolithic target, the greater the deviation from single-plate behavior under high velocity impact conditions. A laminated target is weaker than a solid one of the same thickness. Recall from plate theory that the bending stiffness of a plate is given by

$$D = \frac{Eh^3}{12(1 - \nu^2)}, \qquad (2.26)$$

where h represents plate thickness and ν is Poisson's ratio. If a single-layer plate is cut into, say, two layers, the bending stiffness of each layer is now $(h/2)^3$ or $h/8$. Even though the above comes from consideration of the quasi-static elastic behavior of plates while

64 *Introduction to Hydrocodes*

plate penetration is definitely a dynamic high-rate situation involving significant plasticity, it serves to make the point that layering can weaken a plate significantly.

Aside from the differences in bending stiffness between layered and monolithic plates, adjacent plates of identical impedance that are free to slide transmit compressive waves like their single-plate counterpart. However, rarefaction waves are not transmitted between plates. Thus, deformations and stress propagation characteristics of a multi-plate configuration will differ from that of a monolithic equivalent. Zukas [1996] and Zukas &

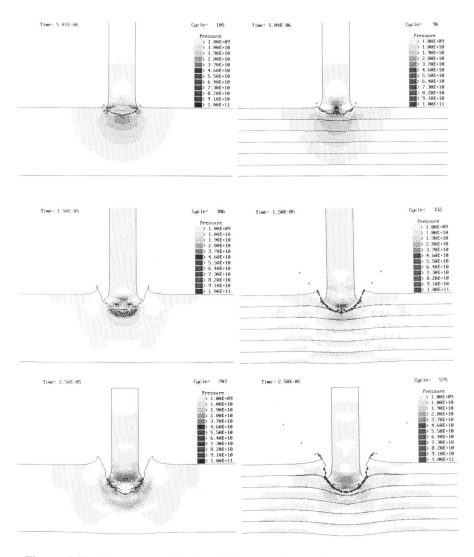

Figure 2.32 Wave propagation in solid and equivalent thickness six-layer plate.

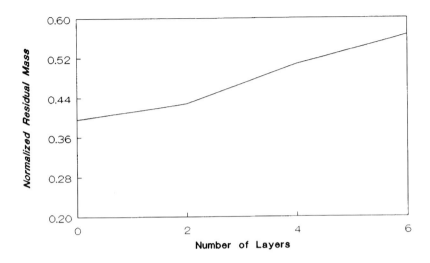

Figure 2.33 Variation of projectile residual mass with target layering.

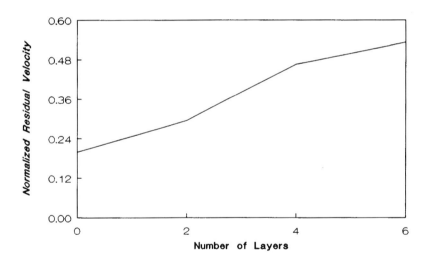

Figure 2.34 Variation of projectile residual velocity with target layering.

Scheffler [2001] studied the effects of lamination and spacing on penetration of monolithic and multi-layered targets. The greatest differences in behavior are observed for thin targets. The behavior of laminated plates and their monolithic counterparts converges with increasing plate thickness. Figure 2.32 shows a comparison of penetration of a solid and six-layer plate of equivalent thickness at various times. Figures 2.33 and 2.34 show the variation of residual mass and velocity with target layering.

66 *Introduction to Hydrocodes*

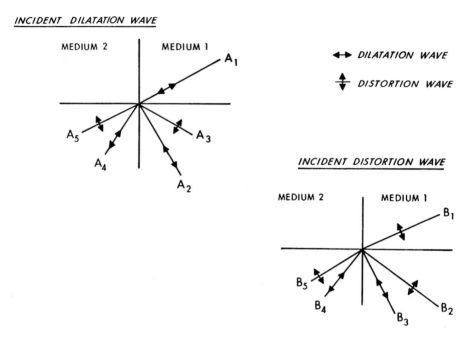

Figure 2.35 Transmission and reflection of incident oblique waves [Rinehart, 1975].

d)   Oblique impact.

Everything we have talked about so far assumes normal incidence of the stress pulse. What if this is not the case? Figure 2.35 shows that when a single pulse impinges on an interface, two waves, one longitudinal, one shear, are transmitted and reflected. This complicates matters considerably insofar as back-of-the-envelope analyses are concerned. The problem of oblique impact is treated by Rinehart [1975] and Johnson [1972] in a very lucid fashion. Many others have written on the topic. In numerical simulations, wave propagation under oblique impacts or nonsymmetric disturbances is handled automatically. Thus we will say no more on this subject until we get to studies of oblique impact situations in a later chapter.

2.6   Dynamic fracture.

If, at a point in a body, the intensity of a tensile stress wave is sufficiently large *and* its duration sufficiently long, a type of failure known as *spallation* can occur. Spall results from the interaction of stress waves with free surfaces or each other. An example of spall

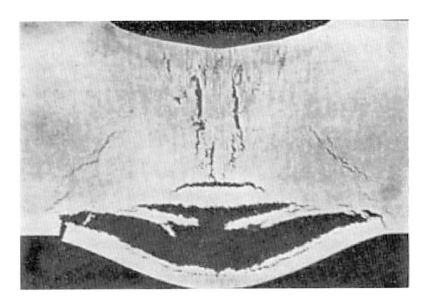

Figure 2.36 Spall failure due to explosive loading [Johnson, 1972].

failure is shown in Fig. 2.36; see Meyers [1994] for a discussion of spallation and models developed for spall failure. Assume that one plate, called the flyer plate, strikes another, referred to as the target plate, at some velocity V. If both plates are of the same material, the particle velocity at the interface will be $v = V/2$ and the stress will be $\rho c v = \rho c V/2$. A compressive stress propagates into both the flyer and target plates. Because the flyer is thinner, its compressive wave will arrive at the rear surface of the flyer plate, be reflected as a tensile wave and return to the flyer–target interface where it will move into the target plate as a rarefaction (relief, tensile) wave. At time $t = C$, the original compressive wave in the target plate reaches the target plate's rear surface and is reflected back into the target as a tensile wave. At a somewhat later time, the tension waves from the flyer and target plates reinforce each other and, if the conditions are right, failure occurs at the spall plane.

Material failure under stress wave loading is very much affected by geometry. Consider Fig. 2.37. Rinehart [1960] has an excellent discussion of geometric effects on wave propagation and failure in brittle materials with many valuable illustrations. Assume that at time t_0 some disturbance has occurred at the center of the plate which causes stress waves to propagate. This could be detonation of an explosive, a projectile impact or some other energetic disturbance. For now, focus only on the wave motion in the plate. Depicted is the primary, or longitudinal, wave. We neglect for this discussion shear waves, which move much slower than longitudinal waves, and we neglect surface waves as well. At time t_1 a spherical wave propagates into the target. At time t_2 the wave has reached the lateral boundaries of our solid and tensile waves reflect back towards

68 *Introduction to Hydrocodes*

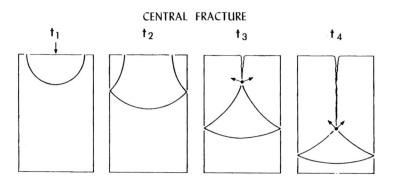

Figure 2.37 Wave reflections leading to central fracture [Rinehart, 1960].

the center while the compressive leading edge of the wave, having encountered no obstacles, continues moving towards the rear of the plate. By time t_3 the two tensile waves reflected from the radial boundaries meet at the centerline, reinforce each other and, *if stress amplitude and duration* are *both* sufficiently large, a crack is formed and propagates down the center of the plate (t_4).

Suppose the disturbance is off-center, at the position indicated by the arrow in Fig. 2.38. Again, a wave starts out from the point of disturbance. Now, however, because the right boundary of the solid is so close, a relief wave is generated at that boundary and follows close on the heels of the initial compressive pulse (t_1). By t_2 the initial compressive pulse reflects from the left boundary and moves to meet with the tensile pulse from the right boundary. Again, if conditions warrant, a crack is formed and propagates down the plate. This is an important aspect of dynamics problems dominated by wave propagation. Failure can occur far away from the point of application of the loading.

Figure 2.39 shows the dynamics of corner fracturing. Figure 2.40 [Rinehart, 1960] summarizes effects of plate dimensions on fracture in a brittle rectangular body by

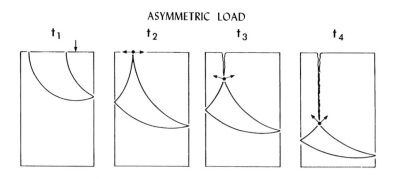

Figure 2.38 Asymmetric fracture due to off-center loading [Rinehart, 1960].

Wave Propagation and Impact   69

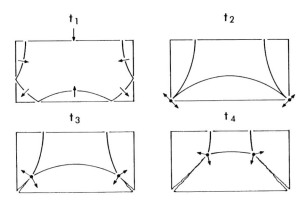

Figure 2.39 Dynamics of corner fracturing [Rinehart, 1960].

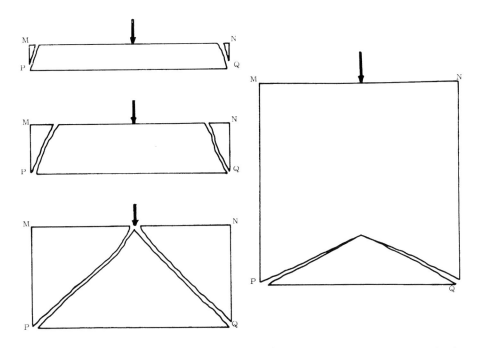

Figure 2.40 Fracture patterns in a rectangular body of varying width-to-height ratios by a propagating spherical wave [Rinehart, 1960].

a propagating spherical wave. The position of the fractures depends on the width-to-height ratio of the rectangle and the geometry of spherical reflection. Many more illustrations of this type, featuring various geometric forms, are in the paper by Rinehart [1960]. Illustrations of some experiments with geometries made of plexiglass are in Johnson [1972].

Plastic wave propagation has been a controversial subject since its inception. Grown men have come to blows and lifelong friendships have been terminated over the debates as to the appropriate theory. Two dominate:

> the *rate-independent* theory proposed by Bell [1960], which holds that there exists a single static curve to describe material behavior at low strain rates and a single dynamic curve to account for high rate behavior.

> the *rate-dependent* theory proposed by Malvern [1951] which allows for a continuous range of strain rates.

Both theories require a general constitutive law. A large number have been proposed. The problem with inelastic wave propagation is that it cannot be neglected, it cannot be solved analytically and, in those few situations where solutions have been obtained, the results have been insensitive to the constitutive equation used.

Modern approaches to the problem involve the use of finite-difference or finite-element methods coupled with an appropriate material model such as the Bodner–Partom Law (see Chapter 4). Uniaxial stress wave propagation experiments involving inelastic behavior are limited to strain rates of several hundreds per second. Recall that 1D analyses for rod-like geometries are limited to elastic behavior and ignore radial (2D effects). This limits the range of stresses and strain rates that can be evaluated. Loads can be increased to drive rods into the plastic range. However, while the experimental observations for such conditions are valid, we are left without analytical means to evaluate the experiments. Thus, we work in a range where strain rates are relatively constant. Wave profiles are not sensitive to the form of the constitutive equation. 3D effects near the impact end of a rod obscure rate effects. 1D analyses are not adequate to resolve these questions. Indeed, they must be validated by 3D analyses.

Perhaps the best description of the problems encountered in inelastic stress wave propagation in uniaxial stress states is given by Nicholas [1975]. If this is a subject close to your heart, you will profit by also reading Bertholf & Karnes [1975], Bodner & Aboudi [1983], Obernhuber, Bodner & Sayir [1986] and Nicholas & Rajendran [1990].

2.7   Summary.

To evaluate computational results for problems involving fast, transient loading we need a basic understanding of: (1) wave mechanics; (2) dynamic material behavior; and (3) the relevant numerical techniques. The first two are associated with the physics of the problem. They serve as a check on the numerical model we create to approximate the physical problem.

From wave mechanics, we need the notion of:

- the wave equation;
- wave speed;
- wave velocity vs. particle velocity;
- stress intensity;
- effects at discontinuities and interfaces;
- irreversible deformations;
- shock waves (uniaxial stress vs. uniaxial strain); and
- the equation of state.

To understand material behavior at high strain rates and how dynamic material data are obtained vs. static or quasi-static data, we need to know:

- phenomenology;
- strain rate effects;
- material failure mechanism; and
- available constitutive models and their limitations.

To analyze dynamic events, we need to know the available tools, be they

- empirical;
- quasi-analytical;
- numerical.

Empirical equations abound and are used daily to compare data, set up experiments and make estimates of impact behavior in related situations. The main danger lies in extrapolation. Stray too far from the database from which parameters for various empirical relationships were determined and disaster can occur. Quasi-analytical methods are a step above the empirical since they at least start with the conservation equations, primarily momentum conservation (a.k.a. as the equation of motion). Complications arise soon enough, however, either because of the complexities of geometry or the nonlinearity of material behavior, or both. Thus, when our knowledge base is exhausted assumptions are made; these can be obvious or subtle. Sooner or later an empirical constant or two is introduced. These are usually presented as "material parameters." Thus, the same caveat applies to these "models" as to purely empirical equations — they are only as good as the database used to determine the constants. If the parameters for a model are determined from experiments with slender fragments striking targets at 500 m/s, do not expect the equations to work for long, slender rods striking targets at 2 km/s. As obvious as this is, it is nothing less than amazing how often equations for impact and penetration are misused in practice. Sometimes hope and a short project deadline triumph over reason. More often than not it is the (apparent) security of having a formula — any formula — and being able to plug numbers into it.

Because the governing equations for dynamic processes are nonlinear rate equations, most practical problems are solved using numerical techniques. These we will address starting in Chapter 4. Thus far, you should have learned that:

(a) Closed form analyses are all 1D.

(b) None take into account the effects of lateral inertia.

(c) There is an unresolved, ongoing debate as to the significance of rate dependence.

So far, everything you have read dealt with idealized situations. The various relationships you have learned are very useful in practice, but do not make the mistake of accepting them as gospel. Examples of the divergence of simple models from reality are given in Chapter 2 of Zukas *et al.* [1982].

REFERENCES

J.D. Achenbach. (1973). *Wave Propagation in Elastic Solids*. Amsterdam: North-Holland.

A. Bedford and D.S. Drumheller. (1994). *Elastic Wave Propagation*. New York: Wiley.

J.F. Bell. (1960). An experimental study of the applicability of the strain rate independent theory for plastic wave propagation in annealed aluminum, copper, magnesium and lead, Technical Report #5, U.S. Army, OOR Contract Number D-36-034-ORD-2366.

L.D. Bertholf and C.H. Karnes. (1975), 2D analysis of the split Hopkinson pressure bar system, *J. Mech. Phys. Solids* (23), 1-19.

R.M. Brach. (1991). *Mechanical Impact Dynamics: Rigid Body Collisions*, New York: Wiley.

S.R. Bodner and J. Aboudi. (1983). Stress wave propagation in rods of elastic–viscoplastic material. *Int. J. Solids Struct.*, 19.

C. Chree. (1886). Longitudinal vibrations of a circular bar. *Quart. J. Pure Appl. Math.*, 21, 287-298.

C. Chree. (1889a). The equations of an isotropic elastic solid in polar and cylindrical coordinates, their solutions and applications. *Trans. Camb. Phil. Soc. Math. Phys. Sci.*, 14, 250-369.

C. Chree. (1889b). On the longitudinal vibrations of aelotropic bars with one axis of symmetry. *Quart. J. Pure Appl. Math.*, 24, 340-359.

D.S. Drumheller. (1998). *Introduction to Wave Propagation in Nonlinear Fluids and Solids*. Cambridge: Cambridge U. Press.

W. Goldsmith. (1960). *Impact*. London: Edward Arnold.

K.F. Graff. (1975). *Wave Motion in Elastic Solids*. New York: Dover.

W. Johnson. (1972). *Impact Strength of Materials*. London: Edward Arnold.

H. Kolsky. (1963). *Stress Waves in Solids*. New York: Dover.

P.H. Netherwood, Jr. (1979). Rate of penetration measurements. U.S. Army Ballistic Research laboratory, Report ARBRL-MR-02978.

I.G. Main. (1993). *Vibrations and Waves in Physics*, 3rd edn. Cambridge: Cambridge U. Press.

L.E. Malvern. (1951). The propagation of longitudinal waves of plastic deformation in a bar of material exhibiting a strain-rate effect. *J. Appl. Mech., Trans. ASME*, 18, 203-208.

M.A. Meyers. (1994). *Dynamic Behavior of Materials*. New York: Wiley.

T. Nicholas and A.M. Rajendran. (1990). Material characterization at high strain rates. In J.A. Zukas, ed, *High Velocity Impact Dynamics*, Chapter 3. New York: Wiley.

T. Nicholas. (1982). Elastic-plastic stress waves. In J.A. Zukas, T. Nicholas, L.B. Greszczuk, H. Swift and D.R. Curran eds, *Impact Dynamics*, New York: Wiley.

K. Nixdorff. (1984). Some applications of the impact theory of J. Awerbuch and S.R. Bodner. *Trans. CSME*, 8(1), 16-20.

P. Obernhuber, S.R. Bodner and M. Sayir. (1986). Numerical analysis of longitudinal elastic–plastic waves with radial effects. *J. Appl. Math. Phys (ZAMP)*, 37.

L. Pochhammer. (1876). Uber die Fortpflanzungsgeschwindigkeiten kleiner Schwingungen in einem unbegrenzten istropen Kreiszylinder. *J. reine angew. Math.*, 81.

A.M. Rajendran and N. Elfer. (1989). Debris-impact protection of space structures. In J. Wierzbicki and N. Jones, eds, *Structural Failure*, New York: Wiley.

J.S. Rinehart. (1960). On fractures caused by explosions and impacts. *Q. Colorado School Mines*, 55(4).

J.S. Rinehart. (1975). *Stress Transients in Solids*. Santa Fe: HyperDynamics.

S.B. Segletes and J.A. Zukas. (1989). The effect of material interfaces on calculations of plate penetration, In D. Hui and N. Jones, eds, *Recent Advances in Impact Dynamics of Engineering Structures*, AMD-vol. 105. New York: ASME.

R. Skalak. (1957). Longitudinal impact of a semih-infinite circular elastic bar. *Trans. ASME, J. Appl. Mech.* 34.

W.J. Stronge. (2000). *Impact Mechanics*. Cambridge: Cambridge U. Press.

J.W. Tedesco and D.W. Landis. (1989). Wave propagation through layered systems. *Comput. Struct.*, 32(3–4).

D.H. Towne. (1967). *Wave Phenomena*. New York: Dover.

F. Whipple. (1952). Meteoric phenomena and meteorites, in *The Physics and Medicine of the Upper Atmosphere*, Albuquerque: U. of New Mexico Press.

R.L. Woodward and I.G. Crouch. (1988). Analysis of the perforation of monolithic and simple laminate aluminium targets as a test of analytical deformation models, Materials Research Laboratory, Report MRL-R-1111, Maribyrong, Victoria, Australia.

J.A. Zukas (ed). (1990). *High Velocity Impact Dynamics*. New York: Wiley.

J.A. Zukas. (1996). Effects of lamination and spacing on finite thickness plate perforation. In N. Jones, C.A. Brebbia and A.J. Watson, eds, *Structures Under Shock and Impact IV*. Southampton: Computational Mechanics Publications.

J.A. Zukas and D.R. Scheffler. (2001). Impact effects in multilayered plates. *Int. J. Solids Struct.*, 38, 3321-3328.

J.A. Zukas and S.B. Segletes. (1991). Hypervelocity impact on space structures, In T.L. Geers and Y.S. Shin, eds, *Dynamic Response of Structures to High-Energy Excitations*, AMD-vol. 127/PVP-vol. 255. New York: ASME.

J.A. Zukas, T. Nicholas, H.F. Swift, L.B. Greszczuk and D.R. Curran. (1982). *Impact Dynamics*. New York: Wiley.

Chapter 3
Shock Waves in Solids

3.1 Introduction.

The bar geometries we considered in Chapter 2 give rise to a state of *uniaxial stress* which is amenable to 1D analysis. If the loading is pushed much beyond the elastic limit, 3D features, such as necking, radial inertia, heating set in, as does material failure. Our simple approximations no longer apply and our analysis fails, although the experimental results are perfectly valid. However, once the uniaxial stress approximation is violated, a different approach is required.

Studies of material behavior involving bar geometries are limited to stresses near the material's yield strength and strain rates between 10^2 and $10^3 \, \text{s}^{-1}$. Typical uniaxial stress–strain curves are shown in Fig. 3.1. It is clear that even with the presence of hardening the stress cannot rise very much. Factors of 2–3 increases in stress above yielding can be obtained in some materials before deviation from a 1D stress state or the onset of failure. The material's plasticity limits us from going higher. Thus to investigate regimes involving higher stresses and strain rates, we must look elsewhere.

Historically, work on shock waves has been done with plate geometries. Plate impact situations generate a state of *uniaxial strain* but 3D stress. Plate geometry offers the opportunity to study material behavior at higher loads and shorter times while offering again the simplicity of a 1D analysis, this time for uniaxial strain. However, just as bar theories neglected lateral inertia, plate impact theories neglect the effects of thermomechanical coupling, which can be significant at strains exceeding 30% [Lee, 1971]. Additionally, much of the initial work assumed hydrodynamic behavior of materials (hence the unfortunate term "hydrocodes" for computer programs dealing with severe impact and explosive loading). However, an elastic precursor can produce significant volumetric strain. An elastic unloading wave can have a strong effect on the local state of the material before the arrival of a plastic wave so that in some cases finite elastic and plastic effects may need to be taken into consideration.

The conventional uniaxial stress–strain curve, as depicted in Fig. 3.1, does not adequately represent the state of stress and strain to which a material is subjected under shock loading. Therefore, the quantities associated with such a curve (elastic modulus, yield strength, ultimate strength and elongation) are not by themselves appropriate to

76 Introduction to Hydrocodes

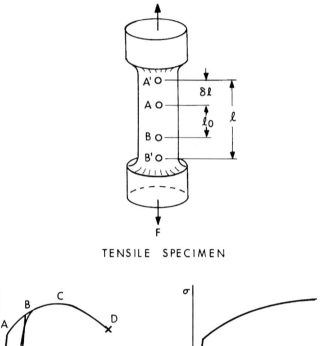

Figure 3.1 Test specimen and typical stress–strain curves for uniaxial stress states.

describe the relative behavior of materials. In order to see how materials behave under shock loading, let us change our perspective.

3.2   Uniaxial strain.

Visualize a situation where deformation is restricted to *one dimension* such as in the case of plane waves propagating through a material where dimensions and constraints are such that the lateral strains are zero, Fig. 3.2. For this situation, the stress–strain curve takes on the form shown in Fig. 3.3.

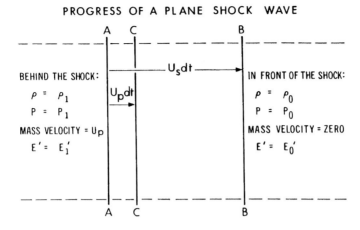

Figure 3.2 Detail of plane shock wave propagating in a solid.

To understand the change from Fig. 3.1 to Fig. 3.3, consider the stresses and strains which occur for 1D deformation. Remember that we are dealing with situations where strains are <30%; else, thermomechanical coupling becomes significant and a more complex analysis is required to interpret experiments.

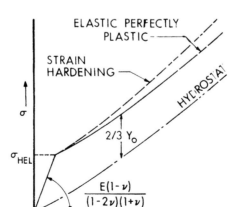

Figure 3.3 Stress–strain curve for uniaxial strain states.

78 Introduction to Hydrocodes

Divide the three principal strains into elastic and plastic components, to wit:

$$\varepsilon_1 = \varepsilon_1^e + \varepsilon_1^p, \tag{3.1}$$

$$\varepsilon_2 = \varepsilon_2^e + \varepsilon_2^p, \tag{3.2}$$

$$\varepsilon_3 = \varepsilon_3^e + \varepsilon_3^p. \tag{3.3}$$

The superscripts e and p refer to elastic and plastic, respectively, and the subscripts are the three principal directions.

In 1D deformation

$$\varepsilon_2 = \varepsilon_3 = 0. \tag{3.4}$$

Therefore

$$\varepsilon_2^p = -\varepsilon_2^e; \quad \varepsilon_3^p = -\varepsilon_3^e. \tag{3.5}$$

The plastic portion of the strain is taken to be incompressible, so that

$$\varepsilon_1^p + \varepsilon_2^p + \varepsilon_3^p = 0. \tag{3.6}$$

With $\varepsilon_2^p = \varepsilon_3^p$ due to symmetry we get

$$\varepsilon_1^p = -\varepsilon_2^p - \varepsilon_3^p = -2\varepsilon_2^p. \tag{3.7}$$

Therefore, keeping in mind that from the principal strain relations with our assumptions $\varepsilon_1^p = 2\varepsilon_2^e$,

$$\varepsilon_1 = \varepsilon_1^e + \varepsilon_1^p = \varepsilon_1^e + 2\varepsilon_2^e. \tag{3.8}$$

The elastic strain in terms of the stresses and elastic constants is given by

$$\varepsilon_1^e = \frac{\sigma_1}{E} - \frac{\nu}{E}(\sigma_2 + \sigma_3) = \frac{\sigma_1}{E} - \frac{2\nu}{E}\sigma_2 \quad (\text{since } \sigma_2 = \sigma_3), \tag{3.9}$$

$$\varepsilon_2^e = \frac{\sigma_2}{E} - \frac{\nu}{E}(\sigma_1 + \sigma_3) = \frac{1-\nu}{E}\sigma_2 - \frac{\nu}{E}\sigma_1, \tag{3.10}$$

$$\varepsilon_3^e = \frac{\sigma_3}{E} - \frac{v}{E}(\sigma_1 + \sigma_2) = \frac{1-v}{E}\sigma_3 - \frac{v}{E}\sigma_1. \tag{3.11}$$

Using the above and the relationship for ε_1 gives

$$\varepsilon_1 = \frac{\sigma_1(1-2v)}{E} + \frac{2\sigma_2(1-2v)}{E}. \tag{3.12}$$

The plasticity condition for either the Tresca or von Mises conditions for this case is

$$\sigma_1 - \sigma_2 = Y_0. \tag{3.13}$$

Using this as the definition for σ_2 gives

$$\sigma_1 = \frac{E}{3(1-2v)}\varepsilon_1 + \frac{2}{3}Y_0 = K\varepsilon_1 + \frac{2Y_0}{3}. \tag{3.14}$$

where $K = E/3(1-2v)$ is called the *bulk modulus*.

For a formal treatment of this topic see Graham [1992], Meyers [1994] and Drumheller [1998] among others. For now observe that if we solve the above equations for stress in terms of pressure, the result would be

$$\sigma_1 = P + \frac{4}{3}Y_0. \tag{3.15}$$

This is the stress–strain relation for uniaxial strain. For uniaxial stress it was $\sigma = E\varepsilon$. Thus, the most important difference between uniaxial stress and uniaxial strain is the *bulk compressibility*. The stress now continues to increase regardless of the yield stress or strain hardening.

For high-rate phenomena where the material does not have time to deform laterally, a condition of uniaxial strain occurs. Later in time, as relief waves arrive from the lateral surfaces and lateral deformations begin to occur, stresses will decrease and a condition approaching uniaxial stress may occur.

For the special case of *elastic* 1D strain

$$\varepsilon_1 = \varepsilon_1^e, \tag{3.16}$$

$$\varepsilon_2 = \varepsilon_2^e = \varepsilon_3 = \varepsilon_3^e = 0, \tag{3.17}$$

$$\varepsilon_1^P = \varepsilon_2^P = \varepsilon_3^P = 0, \tag{3.18}$$

$$\varepsilon_2^e = 0 = \frac{1-v}{E}\sigma_2 - \frac{v}{E}\sigma_1, \tag{3.19}$$

$$\sigma_2 = \left(\frac{v}{1-v}\right)\sigma_1, \tag{3.20}$$

and

$$\varepsilon_1 = \frac{\sigma_1}{E} - 2v^2 \frac{\sigma_1}{E(1-v)}, \tag{3.21}$$

or, finally

$$\sigma_1 = \frac{(1-v)}{(1-2v)(1+v)} E\varepsilon_1. \tag{3.22}$$

Figure 3.4 shows representative stress–strain curves for uniaxial stress and strain states, respectively. There are several notable differences:

(1) There is an increase in modulus for the uniaxial strain curve by a factor of $(1-v)/[(1-2v)(1+v)]$.

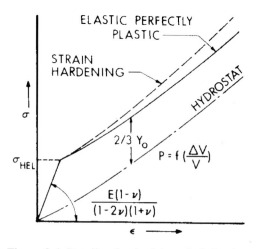

Figure 3.4 Details of uniaxial strain behavior.

(2) The yield point for uniaxial strain is referred to as the Hugoniot Elastic Limit, written as σ_{HEL}. This is the maximum stress for 1D *elastic* wave propagation in plate geometries.

(3) The curve labeled "uniaxial strain" is also known as the *Hugoniot curve*. Note there is a constant deviation from the Hugoniot curve of the stress σ_1 by $2Y_0/3$ where Y_0 is the static yield strength. If the yield strength changes in a strain-hardening material, so will the difference between the σ_1 and P curves. The hydrostat is the curve that a material would follow if it were strengthless.

A typical loading cycle for an elastic-perfectly plastic material in uniaxial strain is shown in Fig. 3.5. Note in particular that reverse loading occurs at point C. If reverse loading occurs, as in stress-wave reflections from a free surface, the line segment CD

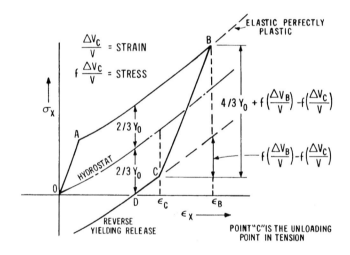

Figure 3.5 Loading–unloading cycle in uniaxial strain.

extends to the negative (tension) region below the strain axis but again different by $2Y_0/3$ from the hydrostat assuming tensile and compressive yield strengths are equal.

One very important point is sign convention. Shock physicists tend to work with compressed states more often than tensile ones. Thus, for convenience, stress and pressure are defined as positive quantities in compression and negative quantities in tension. Structural engineers, on the other hand, are more concerned with tensile behavior of structural components. Thus, to keep from dragging too many minus signs around, they tend to define stress in tension as positive while stress in compression is negative. In dealing with computations involving high-rate loading, we need to use data about materials strength generated by engineers as well as equation of state (EOS) data to represent very high pressures, which are the province of the shock physics community. Thus, it is necessary to pay attention to the sign conventions used by both groups to avoid generating nonsensical results.

3.3   Wave propagation.

Figure 3.6 shows our uniaxial strain stress–strain curve taken to much higher load levels. If the loading does not exceed the Hugoniot elastic limit, a single, elastic wave will

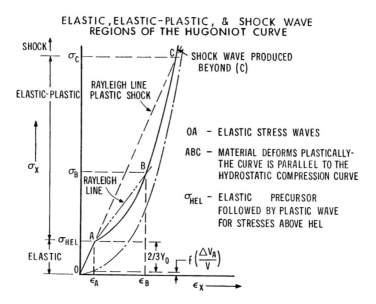

Figure 3.6 Regions of elastic, elasto-plastic and shock wave propagation.

propagate in the material. If the magnitude of the applied stress pulse exceeds σ_{HEL}, two waves will propagate through the medium. The elastic wave will move with a speed

$$c_E^2 = \frac{E(1 - v)}{\rho_0(1 - 2v)(1 - v)}. \tag{3.23}$$

This will be followed by a plastic wave moving with a speed that is a function of the slope of the stress–strain curve *at a given value of strain*. Do not miss the implication of this remark. There can be a multiplicity of plastic waves, each a function of a particular value of plastic strain. The plastic wave speed is given by

$$c_p = \sqrt{\frac{1}{\rho_0} \frac{d\sigma}{d\varepsilon}}. \qquad (3.24)$$

Above the value of σ_0 in Fig. 3.6, we encounter the region of strong shock waves. The material behaves plastically and displays characteristics similar to a fluid. A single, steep-fronted shock wave propagates in this region with a velocity typically denoted as U and determined with the help of an EOS. More on this shortly.

How do we get to this state? In the elastic region, the wave or sound velocity c in the material is a constant. In general, sound velocity is proportional to the ratio of the change in pressure with a change in density

$$c = \frac{dP}{d\rho}. \qquad (3.25)$$

In the elastic region, pressure and density are linearly related. Beyond the elastic region, the wave velocity increases with pressure or density and P/ρ is not linearly proportional. Wave velocity continues to increase with stress or pressure.

Cooper [1996] has written a very lucid and readable description of shock phenomenon. I strongly recommend that this be read before tackling more mathematical treatments. Much of what follows borrows liberally from that text. Figure 3.7 shows

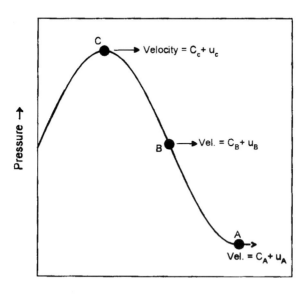

Figure 3.7 Propagating high-pressure wave [Cooper, 1996]. This material is used by permission of John Wiley & Sons, Inc.

84 *Introduction to Hydrocodes*

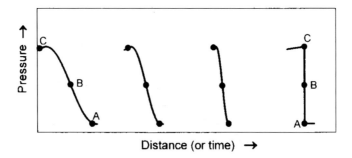

Figure 3.8 Buildup of a pressure wave to a shock wave [Cooper, 1996]. This material is used by permission of John Wiley & Sons, Inc.

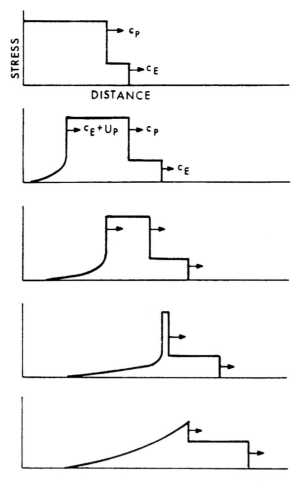

Figure 3.9 Decay of a shock wave due to a rarefaction wave catching up from the rear.

a portion of a pressure wave traveling to the right. At point A, the pressure is low. Therefore, the particle velocity, or speed to which material locally has been accelerated is also quite low. Hence the total velocity of the pressure wave is low. At point B the wave velocity is higher than point A because, above the elastic limit, wave velocity increases with increasing pressure. At point C, wave velocity is higher yet. The net result is shown in Fig. 3.8. The wave continues to get steeper and steeper until it approaches a straight vertical line.

When the wave assumes this vertical front it is called a shock wave. Now there is no longer a smooth transition of matter in front of the wave to matter behind the wave. There is instead a discontinuity between shocked and unshocked material.

If the applied loading is of finite duration, an elastic unloading wave is generated after the removal of the load, Fig. 3.9. The unloading wave travels faster than the compressive wave so that for a short-duration pulse the compressive amplitude may be attenuated by unloading from the rear. The point at which the unloading occurs is called the catch-up distance and is usually defined in terms of the incident pulse thickness.

3.4 Conservation equations under shock loading: the Rankine–Hugoniot jump conditions.

As you just saw, if we have a situation where $c_p > c_E$ conditions have been created for the formation of a steep plastic front. The more rapidly traveling stress components overtake the slower ones. The continuous plastic wave front breaks down and a single discontinuous shock front is formed traveling at a shock velocity U. Across the shock front, there is a discontinuity in stress, density, velocity and internal energy. Because the pressures generated in a solid, which lead to shock wave formation, can exceed material strength by factors of 10–100, it becomes reasonable to consider the solid as behaving like a compressible fluid. Its behavior will be described by an EOS. For now, consider the equation of state, a relationship between pressure, density and internal energy. Think of it as the constitutive model for materials at high pressures.

Shock wave propagation has inherent simplifying features analogous to the case of elastic waves (linear elastic behavior) which permit some simplified solutions for the 1D strain case.

Consider a uniform pressure P_1 suddenly applied at one face of a plate made of compressible material. Assume that the lateral dimensions of the plate are so large in comparison to the dimension in which the disturbance travels that, for all practical purposes, we may consider them to be semi-infinite. Another way of putting it is that we will learn all that we need to learn about material behavior before the arrival of lateral release waves destroys our condition of uniaxial strain. Assume that the material is initially at a pressure P_0. The pressure pulse propagates at a velocity U_s. Application of P_1 compresses the material to a new density ρ_1 and at the same time accelerates the compressed material to a velocity u_p. Now consider a segment of the material (with unit cross-sectional area) normal to the direction of travel. The position of the shock front at some instant of time is indicated in Fig. 3.2 by the line AA. Some short interval of time (dt) later, the shock front has advanced to BB while the matter initially at AA has moved to CC. Across the shock front, mass, momentum and energy are conserved.

Conservation of mass across the shock front may be expressed by noting that the mass of material encompassed by the shock wave $\rho_0 U_s\, dt$ now occupies the volume $(U_s - u_p)dt$ at a density ρ_1

$$\rho_0 U_s = \rho_1 (U_s - u_p). \tag{3.26}$$

Alternatively, in terms of the specific volume $V = \dfrac{1}{\rho}$

$$V_1 U_s = V_0 (U_s - u_p). \tag{3.27}$$

Conservation of momentum is expressed by noting that the rate of change of momentum of a mass of material $\rho_0 U_s\, dt$ in time dt accelerated to a velocity u_p by a net force $P_1 - P_0$ is given by

$$P_1 - P_0 = \rho_0 U_s u_p. \tag{3.28}$$

Conservation of energy across the shock front is obtained by equating the work done by the shock wave with the sum of the increase of both kinetic and internal energy of the system. Thus

$$P_1 u_p = \frac{1}{2} \rho_0 U_s u_p^2 + \rho_0 U_s (E_1 - E_0). \tag{3.29}$$

Eliminating U_s and u_p results in the conservation of energy in the popular form known in the literature as the Rankine–Hugoniot relation

$$E_1 - E_0 = \frac{1}{2}(V_0 - V_1)(P_1 + P_0) = \frac{1}{2}\left(\frac{1}{\rho_0} - \frac{1}{\rho_1}\right)(P_1 + P_0). \tag{3.30}$$

The three conservation equations contain a total of 8 parameters—ρ_0, ρ_1, P_0, P_1, U_s, u_p, E_1, E_0. If it is assumed that the quantities with zero subscripts are known (i.e., the state of the material ahead of the shock), then three equations with five unknowns remain. In order to solve shock problems, we need two more relationships. One such relationship is the ubiquitous EOS. Specification of boundary conditions completes the problem.

3.5   The Hugoniot.

Thus far you have encountered the Rankine–Hugoniot jump conditions and the Rankine–Hugoniot relation. You have also seen a curve on the stress–strain plot for uniaxial strain labeled the "Hugoniot." Confused? I do not blame you.

From now on, when referring to the Hugoniot, we will be talking about a curve determined experimentally. This is also what the literature refers to 99% of the time when mention is made of the Hugoniot, or the Hugoniot equation, with no other qualifiers given. This curve, however, can be displayed in a number of ways. Recall that in the previous section we derived the conservation equations and added to those an EOS. The equation of state gives *all* the equilibrium states in which a material can exist. It is most often written in terms of specific internal energy, pressure and specific volume. For gases we have as the EOS $PV = nRT$ (where v represents volume). For solids, we do not have a general equation of state that can be derived for all materials.

For considering shock transitions only, we do not need all this generality. If we had an EOS in the form $E = f(P, V)$, then it could be combined with the energy jump equation and the energy, E, could be eliminated, giving a relationship of the form $P = f(V)$. This is the Hugoniot equation. In the mass and momentum equations we have two equations in four variables. If we can find a relationship, albeit an experimental one, involving any two of these that also describes material states, then we have an alternative to determining our Hugoniot relation from other than an EOS. Out of the four variables, we could determine a Hugoniot equation relating any two. Our choices would be from among the following [Cooper, 1996]:

$$P-U, \ P-u, \ P-V, \ U-u \text{ and } u-V. \tag{3.31}$$

Our Hugoniot equation then can be determined from experiment. Out of the possible six variable pair planes listed above, three are found to be especially useful: the $U-u$ plane, the $P-V$ plane and the $P-u$ plane.

Before moving on, it is necessary to emphasize that the very nice curves you will see in the next few sections are really fits to experimental data. In uniaxial stress (rod geometry), the stress–strain curve is obtained by a single experiment. For static or quasi-static conditions, an appropriate sample is manufactured to American Society for

88   Introduction to Hydrocodes

Testing Materials (ASTM) standards, inserted into a mechanical testing machine and tested in tension or compression to produce a curve of the type shown in Fig. 3.1. Plate geometry is used to generate Hugoniot data. In a plate configuration, the speed of impact is directly proportional to pressure (remember $P = \rho U u$). Thus, a number of experiments are done at various flyer plate velocities to generate data points in the range of interest. A curve is then fit through the data points; that curve is often referred to as the Hugoniot curve. As with any empirical equation it is only good within the range of parameters that were used to generate the curve. Extrapolation can, and often does, produce nonsense. The shape of the curve depends on how we choose to view our data.

a)   **The U–u plane.**

If we choose to plot our data in the plane of shock velocity vs. particle velocity, the data tend to cluster about a straight line, Fig. 3.10. The equation of that line is often

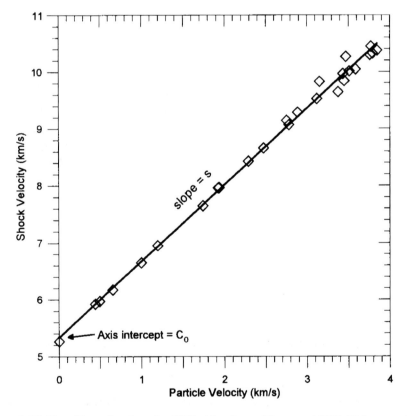

Figure 3.10 U–u Hugoniot data for 6061 Aluminum [Cooper, 1996]. This material is used by permission of John Wiley & Sons, Inc.

given as

$$U = C_0 + su. \qquad (3.32)$$

C_0 is the zero pressure intercept of the curve and s its slope. The quantities C_0, U, u have units of velocity, usually millimeters per microsecond or kilometers per second. The slope s is dimensionless. C_0 is sometimes referred to as the bulk sound speed. For many materials it does approximate the bulk sound speed given by

$$C_B^2 = C_L^2 - \frac{4}{3}C_S^2. \qquad (3.33)$$

where C_L is the elastic longitudinal sound speed and C_S is the transverse sound speed, or shear wave speed. However, it is in reality a parameter derived from a curve fit to experimental data.

Some compilations give the Hugoniot equation in the U–u plane as

$$U = C_0 + su + qu^2. \qquad (3.34)$$

This does not imply that the data fit is not linear. As shown in Fig. 3.11 [Cooper, 1996], the data may be composed of two or more straight line segments or include a phase change.

Compilations of shock Hugoniot data have been published by van Thiel [1977], Marsh [1980], Dobratz & Crawford [1985], and Steinberg [1991].

b) **The P–V plane.**

If we combine the U–u Hugoniot with the mass and momentum conservation equations derived above, we get

$$P = f(V) = C_0^2(V_0 - V)[V_0 - s(V_0 - V)]^{-2}. \qquad (3.35)$$

Look at Fig. 3.12. The Hugoniot represents all the states (P_1, V_1) behind the shock front that can be reached from an initial state in front of the shock (P_0, V_0). The popular description is that the Hugoniot is the locus of all states reachable by a shock transition. It is not, however, the loading path. Do a single flyer plate experiment and material is moved from the "0" state to the "1" state instantaneously. Loading occurs along the *Rayleigh* line. The Rayleigh line represents the jump condition. Its equation can be shown

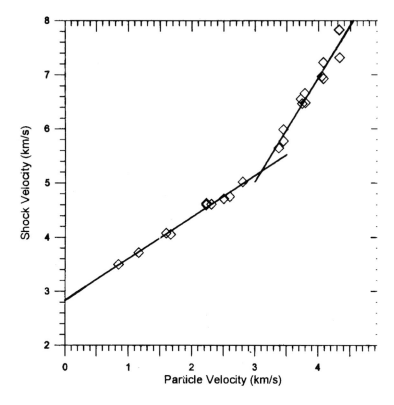

Figure 3.11 $U-u$ data for quartz ceramic [Cooper, 1996]. This material is used by permission of John Wiley & Sons, Inc.

to be

$$P_1 - P_0 = \frac{U^2}{V_0} - \frac{U^2}{V_0^2} V_1. \tag{3.36}$$

Unloading occurs along the isentrope. The isentrope of the material, however, is very close to its Hugoniot so that in solving practical problems we may consider the Hugoniot to be the unloading path. The area between the Rayleigh line and the unloading isentrope, labeled RELIEF in Fig. 3.12, represents the energy retained in a sample undergoing shock loading. This retained energy is responsible for the heating of the sample. Cooper [1996] shows how the temperature increase in the sample may be calculated by knowing how much energy is retained during the shocking process.

Recall once again that the Hugoniot curve is determined from a number of plate impact experiments as shown in Fig. 3.13. Each experiment is conducted at a different loading pressure. This is done by varying the flyer plate velocity in an impact experiment. Loading occurs along the Rayleigh line, which connects individual pressure–volume points with the initial condition.

Shock Waves in Solids 91

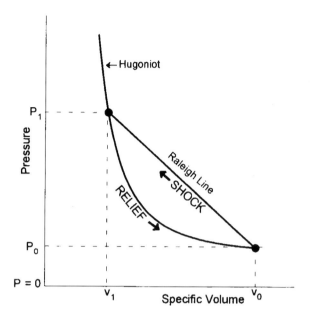

Figure 3.12 Shock loading and unloading [Cooper, 1996]. This material is used by permission at John Wiley & Sons, Inc.

There are several popular representations for the Hugoniot in the $P-V$ plane. Plots of P vs. $\eta(=\rho/\rho_0)$ or P vs. $\mu(=(\rho/\rho_0)-1)$ are often encountered.

c)   **The P–u plane.**

Starting with the $U-u$ Hugoniot and momentum conservation, we can manipulate the equations to eliminate U. With $P_0 = u_0 = 0$ we get

$$P_1 = \rho_0 u_1 (C_0 + s u_1). \tag{3.37}$$

This form of the Hugoniot is displayed in Fig. 3.14. Its primary value is in dealing with shock wave interactions with each other. For details see Cooper [1996].

3.6   The equation of state.

What precisely is an EOS? If you ask a shock physicist, then an EOS will most likely be defined as an attempt to connect (in functional form) the theoretical predictions of microstructural models of atomic lattices to experimental observations of

92 *Introduction to Hydrocodes*

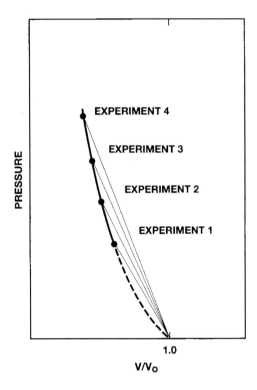

Figure 3.13 Experimental determination of Hugoniot curve [Graham, 1992]. This material is used by permission of John Wiley & Sons, Inc.

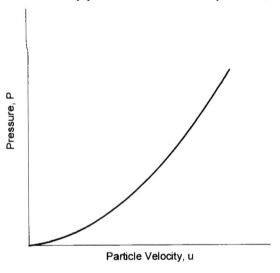

Figure 3.14 Pressure–particle velocity form of the Hugoniot [Cooper, 1996]. This material is used by permission of John Wiley & Sons, Inc.

the macroscopic behavior of metals. No general theory exists which allows us to go from atomic considerations through to prediction of continuum behavior. Thus, simplifications are often introduced. Although interatomic forces in solids result in shear stresses and produce a configuration of triaxial stress during uniaxial deformation, the deviatoric components at equilibrium are ignored. This is equivalent to assuming that the equilibrium stress tensor is spherical and represented by the pressure $P = -\sigma$.

An engineer, by definition someone required to come up with plausible answers to impossible problems in a finite amount of time and funded by a totally inadequate budget, would answer the question somewhat differently. To him, an EOS, would be whatever information about the material is needed that allows us to proceed with whatever problem we are doing (paraphrased from and with apologies to Dr. William C. Davis). Both approaches have been used in computational studies of shock wave behavior. Hydrocodes tend to use something closer to the engineering definition, for the very practical reasons that copious quantities of data are available for a large number of materials and, on the whole, a quasi-empirical approach has produced good results for difficult, practical problems over a span of 50 years [Rinehart (1975)].

For purposes of computations with production hydrocodes, we can say that the EOS is a mathematical description, with some physical basis, of material behavior for a given set of initial conditions. Implicit in any EOS is:

(a) A range of validity of the response. As will shortly be shown, the EOS used in codes is calibrated through plate impact experiments. Current laboratory restrictions limit both the mass and velocity of a flyer plate. Since stress (or pressure) is proportional to flyer plate velocity, a practical upper limit for achievable pressures in metals is in the neighborhood of 6 Mbars. A curve is fit through the experimental data, usually a third-order polynomial, and serves as part of the EOS. It is not difficult to see that using such an EOS on problems where peak pressures of 12–20 Mbars are expected will produce nonsensical results. Because of the experimental calibration, practical EOS formulations are *empirical* relationships, regardless of their theoretical foundations, and all the cautions about not extrapolating beyond the underlying data base apply.

(b) A geometry which makes that loading and response possible. Most EOS data is obtained from normal incidence plate impact geometry. There is a finite observation time beyond which the uniaxial strain state is degraded and simple theories no longer apply. Material failure that is observed is one-dimensional in nature. Plate impact experiments are used primarily to study the compressive behavior of solids, yet the data is applied for the study of geometries where tensile failure may be

94 *Introduction to Hydrocodes*

expected to play an important role. The effects of shear stresses are ignored. These data are then applied to situations involving spherical containers, slender rods and assorted other geometries. The miracle is that it serves as well as it does.

The classic reference for shock waves is the two-volume set by Zel'dovich & Raizer [1967]. Only two chapters are devoted to shocks in solids but these are well worth the time and effort. Drumheller [1998] discusses constitutive equations for elastic, elasto-plastic and shock wave response in a very lucid and thorough manner. Much valuable information on experimental determination of EOS can be found in Meyers [1994] and Asay & Shahinpoor [1993]. It would be difficult to find a better introductory book on the mechanics, physics and chemistry of shock waves than that by Graham [1993]. For an engineering approach to the subject, see Cooper [1996]. An excellent summary of shock waves in gaseous media has been written by Davis [1998]. See the references therein for both the current literature and the classic works in the field. Compilations of EOS data have been published by Kohn [1969], van Thiel [1973], Dobratz [1981], Marsh [1980] and Steinberg [1991].

Let us briefly look at the most widely used EOS formulations in numerical simulations.

a)   *Mie–Gruneisen.*

This formulation is contained in virtually all production hydrocodes. It is used extensively in ordnance velocity (0.5–2.0 km/s) applications. As Drumheller [1998] has pointed out, it is hardly the last word in EOS formulations for solids. However, it has historically had the most attention and, when properly used, it produces results that are good enough for many practical purposes.

The general form of the Mie–Gruneisen EOS is

$$P = P_{\text{ref}} + \Gamma \rho (I - I_{\text{ref}}). \tag{3.38}$$

where P is the pressure; ρ, the density and I, the specific internal energy. Γ is the Gruneisen parameter and is assumed to be a function of volume only. The reference state may be 0 K, in which case I_{ref} is 0 and P_{ref} is the 0 K isotherm. For practical problems such as impact or explosive loading, it is more convenient to take the reference curve as the Hugoniot, so that

$$P_{\text{ref}} = P_{\text{H}}\left(1 - \frac{\Gamma \mu}{2}\right), \tag{3.39}$$

and

$$P_H = C_1\mu + C_2\mu^2 + C_3\mu^3. \tag{3.40}$$

The Cs here are empirical constants determined from plate impact experiments. The quantity $\mu = \rho/\rho_0 - 1$ is a measure of the material's compressibility.

In general, the product of density and the Gruneisen constant is essentially constant. Values of Γ usually lie between 1 and 3. The Gruneisen constant can also be computed from the thermodynamic parameters

$$\Gamma = \frac{3KV\alpha}{C_v}. \tag{3.41}$$

where K is the bulk modulus; $V = 1/\rho$; the specific volume; α, the thermal expansion coefficient and C_v, the specific heat.

Two observations are in order before moving on. The Mie–Gruneisen EOS is only for solids. Thus, if solid–liquid–gas transitions occur in the material during loading or unloading, another EOS must be used. Additionally, most work in determining the shock response of materials has involved pure metals. Consequently, their EOS is well known and catalogued. In practice we encounter alloys and mixtures and messy combinations for which no EOS exists. The theoretical analysis for anything beyond pure materials is currently unavailable. However, the good news is that a number of interpolation methods exist which can be used to get the needed data [Meyers, 1994; Asay & Shahinpoor, 1993]. Also, for many applications, a good first-order approximation for the EOS of an alloy is the EOS of its prime constituent.

b)   *Tillotson.*

The Tillotson [1962] EOS was developed for hypervelocity applications where ultra-high pressures and phase transformations (e.g., melting–vaporization).

Tillotson considers the pressure–volume space as broken into four distinct regions (Figs. 3.15 and 3.16). Region I represents material states obtainable through material transitions less violent than shock transition (e.g., isentropic or isothermal compressions). However, since large pressure states are experimentally obtained through a shock compression process, the Tillotson EOS does not address material states in Region I. In effect, any compressions originating from the ambient state are assumed, by Tillotson, to lie along the Hugoniot.

Regions II and III represent material in its solid state. States in these regions are typically obtained by way of isentropic expansion following an initial shock compression. Region III differs from Region II only in the fact that compressions $\mu = \rho/\rho_0 - 1$ are negative in Region III.

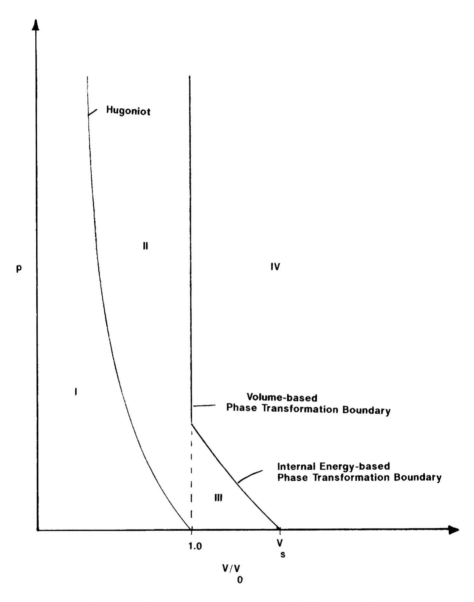

Figure 3.15 Pressure–volume space for Tillotson EOS.

Region IV represents those states where phase transformation to a liquid/vapor has occurred. The boundary between Regions II and IV is the ambient density line, which was chosen for convenience. Between Regions III and IV, the boundary is an iso-energy state where the internal energy is equal to the sublimation energy I_g of the material.

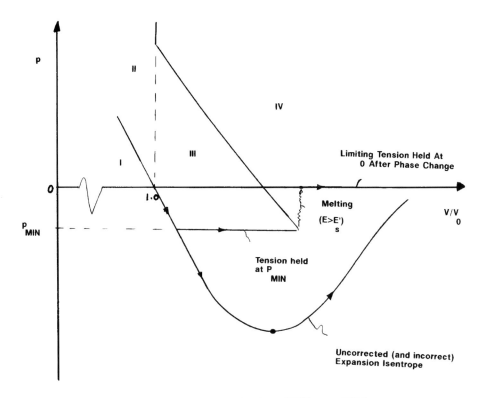

Figure 3.16 Tensile behavior of Tillotson EOS.

The Tillotson EOS is:
REGION II ($\mu > 0$)

$$P = \left[a + \frac{b}{\dfrac{I}{I_0(1+\mu)^2} + 1} \right] \rho_0(1+\mu)I + A\mu + B\mu^2. \qquad (3.42)$$

REGION III ($\mu < 0$ for $I \leq I_s$)

$$P = \left[a + \frac{b}{\dfrac{I}{I_0(1+\mu)^2} + 1} \right] \rho_0(1+\mu)I + A\mu. \qquad (3.43)$$

REGION IV ($\mu < 0$ for $I > I_s$)

$$P = a\rho_0(1+\mu)I + \left[\frac{b\rho_0(1+\mu)I}{\dfrac{I}{I_0(1+\mu)^2}+1} + A\mu e^{-\beta[-\mu/(1+\mu)]}\right] e^{-\alpha[-\mu/(1+\mu)]^2}. \tag{3.44}$$

Here, a, b, A, B, I_0, α, β and I_s are constants with a, b, α, β dimensionless, A and B dimensioned as pressure, I_0, I_s dimensioned as specific internal energy.

Many other forms exist for various applications. See Davis [1998] and Zukas [1990] for examples.

c) **Explosives.**

Most wave propagation codes contain a simple programmed burn model to allow treatment of problems with explosive loading. This is a very simplistic approach, which assumes that an explosive detonates at a given time and the propagating detonation wave transfers its energy to the surrounding materials. To treat the detonation products, a gamma law EOS, $P = (\gamma - 1)(1+\mu)I$ is included. Other EOS for the detonation products may be available as well. Explosive EOS are described by Mader [1979, 1990, 1997]. The performance of an explosive—its ability to push a metal plate or blast a rock—is described by the maximum detonation pressure of the totally decomposed explosive (called the Chapman–Jouguet, or C–J pressure), the detonation velocity and the profile of the pressure of the explosive products as a function of distance and time behind the detonation front. The relationship between pressure, density and temperature of the *explosive products* is called its EOS. The most popular forms encountered in codes are the

1. Becker–Kistiakowsky–Wilson (BKW) EOS

$$\frac{PV}{RT} = 1 + Xe^{\beta X}. \tag{3.45}$$

where P is the pressure, V, the specific volume, R, the gas constant, T, the temperature and $Xe^{\beta X}$, is the imperfection term. The imperfection term is the dominant term and it has a value of 10–15 for most explosives. This EOS is derived and discussed in detail by Mader [1979, 1997].

2. Jones–Wilkins–Lee (JWL) EOS

$$P = A\left(1 - \frac{\omega\eta}{R_1}\right)e^{-R_1/\eta} + B\left(1 - \frac{\omega\eta}{R_2}\right)e^{-R_2/\eta} + \omega\eta I. \qquad (3.46)$$

where A, B, R_1, R_2 and ω are constants to be determined from experiments, I represents internal energy and $\eta = \rho/\rho_0$.

Other forms of EOS for explosives are discussed in the literature. See in particular Asay & Shahinpoor [1993] and Mader [1990, 1997].

3.7 Summary.

From the material above and reading of the literature, one should remember the following:

The study of materials under shock loading requires a change in experimental configuration from rod geometry to plate geometry. In rod geometry (uniaxial stress) the stress intensity is limited by plasticity. In plate geometry (uniaxial strain), the stress intensity is governed by bulk properties—material failure determines the limiting pressure.

Shock waves occur when a material is stressed far beyond its elastic limit by a pressure disturbance.

Because wave velocity increases with pressure above the elastic limit, a smooth pressure disturbance "shocks up."

Because the rarefaction wave moving into the shocked region travels faster than the shock front, the shock is attenuated from behind.

The peak pressure for propagating shocks is given by

$$P = \rho U u. \qquad (3.47)$$

where ρ is the material's density; U, the shock velocity and u, the particle velocity.

Because the pressure generated by shock wave propagation can exceed material strength by several orders of magnitude, we can view the early stages of materials response as hydrodynamic. Inertia governs the process. Strength effects appear in the late stages of the event.

The conservation of mass, momentum and energy equations for transition of unshocked material to a shocked state are known collectively as the Rankine–Hugoniot jump conditions.

The Hugoniot curve is determined from a number of plate impact experiments varying the velocity of the flyer plate for each point on the curve. The loading path is not the curve but the Rayleigh line. The unloading path is the isentrope of the material. For all practical purposes, we take the Hugoniot curve to be the unloading path.

The Hugoniot equation is a fit to data used to generate the Hugoniot curve. Its form depends on the plane in which we choose to view the data. The Hugoniot in the $U-u$ plane for many materials is a straight line given by

$$U = C_0 + su, \tag{3.48}$$

where the parameters C_0 and s are determined from experimental fits to the data. There is no theoretical support for the linearity of this equation. It just works out that way for many materials. For the exceptions to the rule, a higher-order polynomial is used to fit the data.

The Hugoniot is by itself not enough to fully characterize a material. It only describes states attainable by a shock transition. For a complete description we need to add an EOS and initial conditions.

An EOS is an attempt to describe material behavior at the continuum level beginning by considering the interatomic forces and their effects on the lattice structure to a given set of initial conditions. A full description of material behavior of this type cannot at present be obtained from first principles. Thus, EOS used in practice combine experimental data such as the Hugoniot with an underlying theoretical structure. A great number of EOS formulations are available, but only a few are used in wave propagation codes. For impact velocities below 2 km/s where the material remains a solid, the *Mie–Gruneisen* form is very popular, due mainly to the simplicity of the formulation and the availability of data in many compilations. For impacts and explosive loading at higher velocities, where melting and vaporization can occur, the *Tillotson Equation* has been used extensively. For explosive detonation products, the BKW, JWL and gamma law EOS are popular.

Because experimental data are introduced into the EOS, mainly through the Hugoniot, it exhibits the same characteristics as any empirical equation. Extrapolation very far from the underlying data base generates nonsensical results.

EOS data is generated almost always in compression. The EOS is then used to describe tensile behavior in geometries far removed from those for which the data were generated.

If the material in this and the preceding chapter is a complete mystery to you, then you need to spend a year or two studying stress and shock wave propagation in solids subjected to intense, impulsive loading. At this point in your development you should not be using a hydrocode except under the close supervision of an experienced user. You are a danger to yourself and to those who believe the results you generate.

REFERENCES

J.R. Asay and M. Shahinpoor. (1993). *High-Pressure Shock Compression of Solids.* New York: Springer-Verlag.

P.W. Cooper. (1996). *Explosives Engineering.* New York: VCH-Wiley.

W.C. Davis. (1998). Shock waves; rarefaction waves; equations of state. In J.A. Zukas and W.P. Walters, eds, *Explosive Effects and Applications.* New York: Springer-Verlag.

B.M. Dobratz. (1981). LLNL explosives handbook, Lawrence Livermore National Laboratory, Report UCRL-52997.

B.M. Dobratz and P.C. Crawford. (1985). LLNL explosives handbook, Lawrence Livermore National Laboratory, UCRL-52997, Change 2.

D.S. Drumheller. (1998). *Introduction to Wave Propagation in Nonlinear Fluids and Solids.* Cambridge: Cambridge U. Press.

R.A. Graham. (1992). *Solids Under High-Pressure Shock Compression.* New York: Springer-Verlag.

R.A. Graham. (1993). *Solids Under High-Pressure Shock Compression.* NY: Springer-Verlag.

B.J. Kohn. (1969). Compilation of hugoniot equations of state, Air Force Weapons Laboratory, Report AFWL-TR-69-38 (AD852300).

E.H. Lee. (1971). In J.J. Burke and V. Weiss, eds, *Shock Waves and the Mechanical Properties of Solids.* Syracuse, New York: Syracuse U. Press.

C.L. Mader. (1979). *Numerical Modeling of Detonation.* Berkeley, CA: U. of California Press.

C.L. Mader. (1990). Introduction to energetic materials. In J.A. Zukas, ed, *High Velocity Impact Dynamics.* New York: Wiley.

C.L. Mader. (1997). *Numerical Modeling of Explosives and Propellants.* Boca Raton, FL: CRC Press.

S.P. Marsh (ed). (1980). *LASL Shock Hugoniot Data.* Berkeley: U. of California Press, CA.

M.A. Meyers. (1994). *Dynamic Behavior of Materials.* New York: Wiley.

J.S. Rinehart. (1975). *Stress Transients in Solids.* Santa Fe, NM: HyperDynamics.

D.J. Steinberg. (1991). Equation of state and strength properties of selected materials, Lawrence Livermore National Laboratory, Report UCRL-MA-106439, Livermore, CA.

J.H. Tillotson. (1962). Metallic equations of state for hypervelocity impact, General Atomic, Report GA-3216.

M. Van Thiel. (1977). Lawrence Livermore Laboratory, Report UCRL-50108, Vols 1–3.

Y.B. Zel'dovich and Y.P. Raizer. (1967). *Physics of Shock Waves and High-Temperature Hydrodynamic Phenomena*. New York: Academic Press.

J.A. Zukas. (1990). *High Velocity Impact Dynamics*. New York: Wiley.

Chapter 4
Introduction to Numerical Modeling of Fast, Transient Phenomena

4.1 Introduction.

There are four approaches to the solution of nonlinear problems involving fast, transient loading:

Experimental: many problems do not lend themselves to analysis. The only recourse then is testing. Testing can be both time-consuming and expensive. A number of things can go wrong in a test program so that several tests may need to be conducted to obtain a single, valid data point. For example, laboratory-scale impact experiments can be conducted for a few thousand dollars each if only go no go information is required. A ballistic limit, the minimum velocity needed to perforate a target of a given thickness, may be obtained in this fashion. If, however, there is to be additional exploitation of the information obtainable in an impact test, then additional complexity and instrumentation are introduced at a concomitant increase in cost. Thus, to determine not only whether a target was perforated but also to determine residual mass, velocity and orientation of the striker, the hole profile in the target and the characteristics (mass, velocity and orientation) of the debris behind the target, the cost of testing can go to $20,000 per experiment or higher. If testing needs to be performed at full scale, costs may be an order of magnitude higher. Now cranes must be used to move targets weighting tons into position, guns aligned with lasers. Instrumentation must work in all types of weather. This increase in complexity comes with an increase in manpower requirements and cost.

Cost is not an insignificant consideration. In designing a test program one would ideally wind up with a statistically significant data set. This rarely happens. Thus, the available data must be exploited to the greatest degree possible and, if possible, supplemented with analytical and numerical models. Empirical curve fits are developed to existing data sets and these are used extensively to guide further experimentation and design. The drawback to empirical fits is that they cannot be extrapolated. Experimental methods also fail to account for material properties save in the grossest sense—density and strength. The latter is usually given as a hardness number on one of the several popular scales and thus is a measure of surface strength only.

Semi-empirical data fits: when considerable data are available from several test programs, empirical correlation can be attempted on a broader scale, to include variables momentum, mass, velocity, striker or target dimensions, etc. An example of this approach is the THOR equations for fragment penetration [Zukas, 1990]. These are based on a correlation of non-dimensional parameters (which might be determined from intuition, similitude analyses or other considerations) with a large body of data. It is possible to explicitly include simple material properties such as yield strength. The advantage of this approach is that it helps to flesh out an existing database and reduces cost. A disadvantage of any empirical approach is that extrapolation is extremely dangerous and can swiftly lead to erroneous results. If additional data is later added to the existing database, then the empirical relations must be recalibrated.

Engineering models: this approach involves starting with first principles. Conservation of mass, momentum and energy are invoked. Additional simplifications (e.g., 1D behavior, consideration of early, steady-state or late-time effects) may be introduced at the outset or incorporated into the analysis as needed. It may be necessary to assume a form of the solution, e.g., wave propagation, deformation patterns, flow fields. One may also incorporate several material properties such as a yield model, hardening effects, failure criteria, wave speeds or a particular equation of state (EOS) if very high pressures are involved. These models tend to be more useful than empirical relationships because they are more general. However, because of the complexity involved, they may apply only to a portion of the response spectrum. Also, it is not unusual for model developers, when stuck, to postulate material constants which are not easily determined in the laboratory and are, in effect, empirical parameters. These may allow model developer to obtain a closed-form solution (or one easily tractable on a portable computer) but this is done at the expense of generality and the usual caveat about extrapolating empirical models applies.

Numerical models: the most general approach is a numerical one. Its main advantage is that any limitations on the geometry of a problem are removed. Nonlinearities may be taken into account. The drawback is the usual one when proceeding from the simple and specific to the general and complex—more data are required. Numerical models require a complete description of the material behavior in all loading regimes—elastic behavior, yielding, flow, failure. Pressure, strain, strain rate and temperature effects need to be accounted for. This lack of a complete material description, especially as it relates to material failure, is the single greatest limitation on the use of wave propagation computer codes to predict high rate material behavior.

With the limited work now going on in development of 2D models for dynamic events, it is likely that in the near term we will have to rely on experiments, empirical models, back-of-the-envelope estimates and wave propagation codes. If analytical models are available for your situation, by all means use them, taking the usual precaution of not extrapolating too far beyond the database from which model constants are derived. Sooner or later, though, situations will arise where computations with wave propagation codes are the only alternative, e.g., simulating nuclear weapons effects during the life of a test ban treaty. Interestingly, sometimes the reverse is true. Situations regularly arise where computations are possible, but because of excessive run times, sacrifices in accuracy in setting up a grid or gross uncertainties in material behavior, it is faster, cheaper and more informative to do a few experiments than a single calculation.

It will become clear as we go along that computational mechanics is not a hard science (not in the sense of difficult but rather a mixture of art and science), insofar as high strain rate behavior is concerned. Success in this field depends very much on the familiarity of the practitioner with the problem area in which he/she works. Equally important is the choice of computational tool. Each code solves essentially the same set of governing equations. Each one, however, is the unique creation of an individual or a team, has its own quirks and eccentricities and cannot be understood just by reading the users' manual. Considerable experience is required before a code can be used effectively. Three individuals of varying background could use the same code on the same problem and obtain three dramatically different sets of results. In no way can the current crop of wave propagation codes be considered as *black boxes*. Be wary of those who suggest otherwise. In a previous life they probably sold snake oil!

As you embark on a field of study that is as much art as science, keep a keen watch on your frustration level. The more experience you gain, not just in computational techniques but in related areas of structural dynamics, impact dynamics, material behavior at high strain rates, dynamic material failure, experimental techniques (especially in establishing the validity of experimental data) and explosive behavior, the better results you will achieve. This takes time. You may be frustrated by pressures at work or being prematurely elevated to the status of expert by the fiat of an executive committee far out of touch with reality. Persist! The time will come when you will be able to look at a computational result and know what type of code was used to obtain it, the expertise of those who obtained it and what it is worth. For now, let us begin at the beginning.

4.2 Spatial discretization.

All existing structural dynamics and wave propagation codes obtain solutions to the differential equations (DEs) governing the field by solving an analogous set of algebraic equations. The governing DEs are not solved directly. Why? Because currently only a handful of closed-form solutions for DEs are available. The equations of structural dynamics, being a coupled set of rate equations which account for the effects of severe gradients in stress, strain and deformation, material behavior ranging from solid to fluid to gas, temperatures from room temperature to melt temperature are highly nonlinear and do not lend themselves to closed-form solutions in the general case. Try this experiment next time you are in a decent library. Look at the number of books on the shelves, which deal with partial differential equations (PDEs). You will find a few, and these will tend to concentrate on the wave equation and the heat equation. Look for books on nonlinear PDEs and the collection shrinks considerably. If you check out books on ordinary DEs, the size of the collection increases markedly. Continue on, searching for books dealing with algebraic equations and you will find at least a roomful. This says that we have at our disposal a large number of techniques for solving algebraic equations, but only a very few for solving PDEs. Is it possible then to transform our PDEs to algebraic equations and take advantage of the many solution techniques available to us? Very often, yes!

Figure 4.1 shows how we will proceed. First, we will look at the equations we are trying to solve. Next, we will see about transforming these to algebraic equations. Along the way we will have to say a few words about the framework in which we cast these equations—Lagrangian, Eulerian or some hybrid form. In this book the emphasis is on

Numerical Simulation of Fast, Transient Loading

- **Differential Equations**

- **Spatial Discretization**
  Finite Difference Methods
  Finite Element Methods

- **Lagrangian Mesh Description**
  Mesh Characteristics
  Sliding Interfaces
  Large Distortions: rezoning, erosion
  Artificial Viscosity

- **Time Integration**
  Explicit Methods
  Implicit Methods

- **Material Models**
  Incremental-Elastic-Plastic
  Johnson - Cook
  Zerilli - Armstrong
  Bodner - Partom
  $p - \alpha$
  CAP

- **Failure Models**
  instantaneous
  time-dependent
  micromechanical

- **High Strain Rate Material Data**
  Experimental Procedures
  Sources of High Rate Data

Figure 4.1 Transition from differential equations to a hydrocode.

Lagrangian mesh descriptions. Having constructed a set of algebraic equations to cover our spatial domain, we will need to do the same thing in time. Finally, we add to this, descriptors of material behavior at high strain rate and we have constructed an algebraic analog to our original PDEs which, we fervently hope, will behave in the limit in the same way as our original PDEs. Granted, our original set of PDEs could be written down on the back of an envelope whereas we now have to solve thousands, maybe even hundreds of thousands, of coupled algebraic equations, depending on the accuracy we want. But at least we are dealing with an equation set which we can solve with standard techniques as opposed to our original PDEs which we could admire and discuss but not solve.

You will see this material twice. Think of this chapter as "Highlights From Hamlet." The fundamental ideas that go into the creation of a Lagrangian computer code are highlighted here. They are presented, however, in technical shorthand. If you are sufficiently clever, you can proceed to write your own code after reading this chapter. If you did, you would soon have questions about what was presented here as well as others about material that was implied but not explicitly stated. For this reason this chapter, which presents the material as you can expect to find it in the literature, is followed by Chapter 5 which presents it in the form in which it is used in an actual code.

Figure 4.2 lists the conservation equations for solids with shocks in their most general form. This is always a good place to begin. There is something fundamentally unsettling for systems, which do not satisfy energy and momentum conservation.

CONSERVATION EQUATIONS

CONSERVATION OF MASS:
$$\dot{\rho} + \rho \dot{u}_{i,i} = 0$$

CONSERVATION OF LINEAR MOMENTUM:
$$\rho \ddot{u}_i = \sigma_{ji,j} + \rho f_i$$

CONSERVATION OF ANGULAR MOMENTUM:
(a) POLAR MEDIA
$$\rho \dot{L}_i = \rho Q_i + R_{ji,j} + \epsilon_{jk\ell} \sigma_{jk}$$
(b) NON-POLAR MEDIA
$$\sigma_{ij} = \sigma_{ji}$$

CONSERVATION OF ENERGY:
$$\rho \dot{E} = (\sigma_{ji} \dot{u}_i)_{,j} - q_{i,i} + \rho S + \rho \dot{u}_i f_i$$
$$E = I + 1/2 \, \dot{u}_j \dot{u}_j$$

HUGONIOT JUMP CONDITIONS:
$$\rho_0 U_s = \rho (U_s - u_p)$$
$$P - P_0 = \rho_0 U_s u_p$$
$$I - I_0 = 1/2 \, (P + P_0) \left(\frac{1}{\rho_0} - \frac{1}{\rho} \right)$$

Figure 4.2 Conservation equations.

Lagrangian codes satisfy mass conservation by default, so an explicit statement of that equation is not needed. The material models in existing production codes, both Euler and Lagrange, assume nonpolar media. The theory for polar media is well developed, however, and can easily be incorporated. The creation and annihilation of shock waves are also not tracked explicitly in production codes so that the Hugoniot jump conditions are not included. Codes have been developed which explicitly tracked shocks. These have worked well in 1D applications but the so-called shock fitting logic has been difficult to implement in 2D and 3D codes and also adds considerably to the cost of computing. Thus, this feature is not included in production 2D and 3D codes.

What are "production codes?" That phrase refers to codes which are reasonably well documented and intended for use beyond the group or organization that developed the code. A somewhat dated list appears in the appendices to Chapter 9 in Zukas [1990]. In addition to production codes, there are codes proprietary to a specific group or organization, which were developed and intended to be used strictly as in-house tools. As a rule they lack the user-friendly features of production codes since they were intended for use by their developers and close associates and may have little or no documentation.

Figures 4.3 and 4.4 outline some of the more popular material models. More will be said on this topic later. The usual formulation is an incremental elastic–plastic formulation which allows for the effects of strain, strain rate, compressibility and

CONSTITUTIVE EQUATIONS
ELASTIC-PLASTIC MATERIAL:
(a) STRESS DEVIATORS
$$\dot{s}_{ij} = 2G(\dot{\epsilon}_{ij} - \tfrac{1}{3}\delta_{ij}\dot{u}_{k,k})$$
(b) VELOCITY STRAINS
$$\dot{\epsilon}_{ij} = \tfrac{1}{2}(\dot{u}_{i,j} + \dot{u}_{j,i})$$
$$\dot{\omega}_{ij} = \tfrac{1}{2}(\dot{u}_{i,j} - \dot{u}_{j,i})$$
(c) JAUMANN STRESS RATE
$$\hat{\sigma}_{ij} = \sigma_{ij} + \sigma_{im}\omega_{mj} - \omega_{im}\sigma_{mj}$$
(d) TOTAL STRESS
$$\sigma_{ij} = s_{ij} - \delta_{ij} P$$
$$s_{ii} = 0 \qquad \sigma_{ii} = -3P(\rho, 1)$$
(e) VON MISES YIELD CRITERION
$$s_{ij}s_{ij} \leq 2/3\, Y^2$$
IF YIELD STRESS EXCEEDED:
$$s_{ij} = s_{ij}\left[\frac{2Y^2}{3 s_{ij} s_{ij}}\right]^{1/2}$$

Figure 4.3 Constitutive equations for elasto-plastic materials.

CONSTITUTIVE EQUATIONS

FLUID FLOW:

$$\dot{d}_{ij} = \dot{\epsilon}_{ij} - \frac{1}{3}\dot{\epsilon}_{kk}\delta_{ij}$$

$$\dot{d}_{kk} = 0$$

$$\dot{s}_{ij} = c_{ijk\ell}\dot{d}_{k\ell} \quad \text{(ANISOTROPIC)}$$

$$\dot{s}_{ij} = 2G\dot{d}_{ij} \quad \text{(HOMOGENEOUS, ISOTROPIC)}$$

Figure 4.4 Constitutive relations for fluid flow.

temperature (alternatively, internal energy). These follow the formulation originally laid down by Mark Wilkins of Lawrence Livermore National Laboratory in his formulation of the hydrodynamic elastic magneto-plastic (HEMP) code in 1964 [Wilkins, 1999]. Anisotropic homogeneous material behavior is assumed in these codes. However, this does not include the behavior of many composite systems since most are inhomogeneous as well as anisotropic.

High pressures—orders of magnitude above the material strength—are dealt with through an EOS. Representative examples for solids and explosives are shown in Figs. 4.5 and 4.6. By and large the Mie–Gruneisen form is the most popular and has found its way into all production codes. Mie–Gruneisen assumes, however, that a solid remains a solid forever; so for all practical purposes its utility is limited to impact velocities up to about 6 km/s. Where did that number come from? Most compilations of EOS data have information that has been experimentally validated for pressures between 2 and 4 Mbars (2–4×10^{12} dynes/cm^2). For practical materials such as aluminum, steel and a few high

EQUATIONS OF STATE

MIE - GRÜNEISEN:

$$P = P_H\left(1 - \frac{\Gamma\mu}{2}\right) + \Gamma\rho(I - I_0)$$

$$\mu = \rho/\rho_0 - 1$$

TILLOTSON:

FOR $\rho > \rho_0$ AND $0 \leq I \leq I_s$

$$P = P_\pi + A\mu + B\mu^2$$

$$P_\pi = I\rho\left[a + \frac{b}{I/(I_0\eta^2)+1}\right]$$

FOR $\rho < \rho_0$ WITH $I > I_s$

$$P = aI\rho\left[\frac{bI\rho}{(1+I/(I_0\eta^2))} + A\mu e^{-\beta(1/\eta-1)}\right]e^{-\alpha(1/\eta-1)^2}$$

$$\eta = \rho/\rho_0$$

Figure 4.5 Representative equations of state for solids.

EQUATIONS OF STATE

LOS ALAMOS:

$$P = \begin{cases} [A\mu + \rho_0 I(B + \rho_0 IC)]/(\rho_0 I + \phi_0) & \text{IF } \mu \geq 0 \\ [\mu A_1 + \rho_0 I(B_0 + \mu B_1 + \rho_0 IC)]/(\rho_0 I + \phi_0) & \text{IF } \mu \leq 0 \end{cases}$$

WILKINS:

$$P = a\eta^\alpha + b\left(1 - \frac{\omega}{R}\eta\right)e^{-R/\eta} + \omega\eta I$$

GAMMA LAW:

$$P = (\gamma - 1)(1 + \mu)I$$

JONES, WILKINS, LEE (JWL):

$$P = A\left(1 - \frac{\omega\eta}{R_1}\right)e^{-R_1/\eta} + B\left(1 - \frac{\omega\eta}{R_2}\right)e^{-R_2/\eta} + \omega\eta I$$

Figure 4.6 Additional examples of equations of state.

density alloys, this implies striking velocities of about 6 km/s (remember ρcv and ρUu?). If it is known that the solid undergoes a transition to liquid or even vapor, another EOS must be used. Among the most popular for hypervelocity work is that due to Tillotson. Almost all codes have a gamma law EOS for gases. Very popular for explosives is the Jones, Wilkins, Lee (JWL) EOS if you are of the Livermore persuasion or the Becker–Kistiakowski–Wilson (BKW) for the adherents of the Los Alamos school of explosive behavior. Grown men sometimes descend into childish arguments over the merits of these EOS, neither side having the experimental–numerical correlations to carry the day. Enjoy the fun if you are around when it happens but do not take it seriously. Both are fits to empirical data and are no better than the database from which the constants are derived. If you are going to get a bloody nose over something, make it blondes, brunettes or your favorite brand of single-malt whiskey, anything but six-parameter curve fits. Remember Cauchy's claim that given four free parameters he could draw a credible version of an elephant and given five he could make its tail wiggle!

Figure 4.7 recaps one of our problems. Should we use finite differences or finite elements to construct our algebraic analog? Both have been used successfully. Initially (early 1960s) the finite-difference method was used exclusively. In fact, all of today's Euler codes use finite differences. There are many good reasons to use finite differences but the early authors such as Wilkins (HEMP), Herrmann (TOODY), Wally Johnson (OIL, DORF), Walsh and Hageman (HELP), Thompson (CSQ), Matuska (HULL) used them because finite elements had not been invented yet! Gordon Johnson (EPIC) was the first to use finite elements in a wave propagation code, preceding the release of Hallquist's DYNA by a year. EPIC used triangular elements, DYNA used quadrilaterals and this difference was supposedly the reason for some very acrimonious articles published in the open literature denouncing EPIC in general and triangular elements in

DILEMMA

A general solution to coupled, nonlinear partial differential equations (PDE's) cannot be obtained.

However, numerous techniques exist for solving algebraic equations.

Therefore, construct an analogous set of algebraic equations with the same behavior, in the limit, as the original PDE's using:

- finite differences
- finite elements

Figure 4.7 Dilemma.

particular. In my opinion, the actual reason for the poison pen papers is not hard to guess since EPIC and DYNA were able to do the same class of problems with the same accuracy and initially EPIC was by far the more popular code. It was not until coding for specific graphics processors available at Livermore and few other places was removed and replaced with general purpose post-processing routines that the fortunes of DYNA began to grow. Other codes followed so that today most production Lagrange codes use finite elements whereas most production Euler codes use finite differences. The reasons are primarily historical.

a)   *Finite differences.*

Finite-difference methods are sometimes referred to as *approximate* solutions to *exact* problems. The reason for this is that with finite differences we work with the physical relationships governing the processes we wish to study. We manipulate these into DEs (the *exact* part) and then systematically replace derivatives with difference expressions in the DEs, boundary conditions and initial conditions. We then solve the resulting algebraic equations to whatever degree of accuracy we can afford (the *approximate* part) with standard techniques.

Glance at Fig. 4.8. This depicts some portion of a complex, nonlinear curve $y(x)$. We have zoomed in on only a small segment of it and we now wish to approximate its behavior in the interval between x_1 and x_2 by a straight line. We are guaranteed (which theorem of calculus does this? If you cannot remember, see Fried [1979], one of the best and most readable books on the subject; if it is not in your library it should be) that our

112 *Introduction to Hydrocodes*

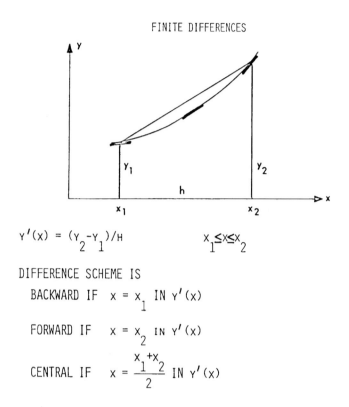

Figure 4.8 Construction of a finite-difference scheme.

approximation will be exact at minimum at one point in the interval. We can now write that our derivative, $y'(x)$, in our interval will be that given in Fig. 4.8. Our difference scheme has different convergence properties depending on where we choose to evaluate our derivative. It is called a *backward scheme* if we choose $x = x_1$, forward if we pick x_2 and central if we evaluate our derivative in the middle of the interval. It can be shown, and this is done very nicely by Fried, that central difference schemes are second-order accurate, or $O(h^2)$. This means that as our interval h is reduced by a factor of 2, our error decreases by a factor of 4—a very nice property indeed. Forward and backward finite-difference schemes are only first-order accurate—$O(h)$.

Figure 4.9 shows how second, and by implication, higher-order derivatives are found. Having constructed such approximations to all of the derivatives in our DE set, we now go back and substitute the difference expressions for the derivatives and evaluate them at as many points as we need (more often, can afford) to obtain a convergent solution to our problem. If you have not done this before it may be a bit hard to visualize, so let us do it for a very practical problem of a simply supported elastic beam subjected to a uniform distributed load w. Figure 4.10 shows the governing DE from strength of

Introduction to Numerical Modeling of Fast, Transient Phenomena 113

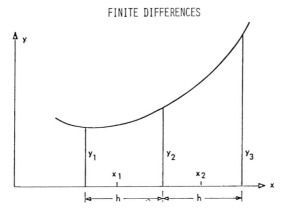

Figure 4.9 Creating higher-order differences.

materials or elasticity theory, together with the appropriate boundary conditions at each end. We can solve the problem as now stated with boundary conditions at $x = 0$ and $x = L$. However, we observe the existence of symmetry in this problem and jump to take advantage of it. Now we need consider only half the beam with new boundary conditions that the deflection at $x = 0$ is 0 and the slope at the midpoint is 0 also.

We begin discretizing (Fig. 4.11) by dividing our region (half the beam, length $L/2$) into N equal sections. How big should N be? That is the eternal question of the numerical analyst. Picking an appropriate N comes with a knowledge of the problem being solved and lots of experience. There are no theorems, which say N should be such and such. Too small an N will not give us the correct solution. Too large an N will give us the solution but it may cost a fortune. We need an N that satisfies both mathematical accuracy and organizational economics. Also, it should not exceed the memory capacity of our computer.

You can arrive at an acceptable N following this prescription. For this particular problem we have a closed-form solution, so we know what the deflection should be.

For more violent problems we may have strain gage records, pressure–time histories, steady-state solutions, at worst post-mortem data (crater dimensions for

114 *Introduction to Hydrocodes*

FINITE DIFFERENCES
EXAMPLE: UNIFORMLY LOADED BEAM

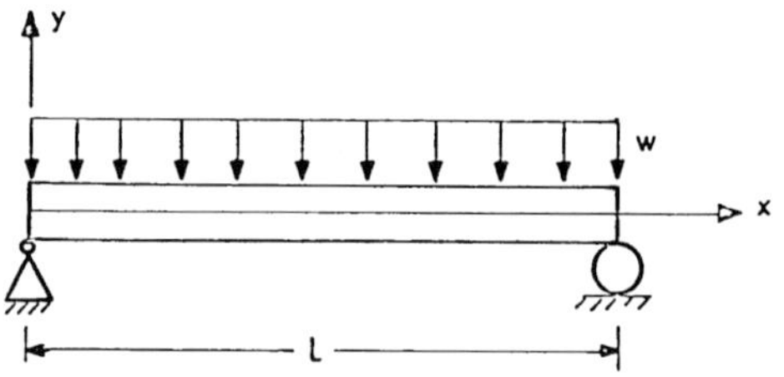

D.E.: $EIy'' = \dfrac{WLx}{2} - \dfrac{Wx^2}{2}$

E = ELASTIC MODULUS
I = MOMENT OF INERTIA
y = DEFLECTION

B.C.: $y(0) = y(L) = 0$

BECAUSE OF SYMMETRY, CONSIDER ONLY HALF OF BEAM

$0 \leq x \leq L/2$ WITH B.C. $y(0) = 0$

$y'(L/2) = 0$

Figure 4.10 Simply supported uniformly loaded beam.

semi-infinite targets; residual masses, velocities and orientation of the penetrator and the hole profile in a perforated finite-thickness plate). Pick an N! The better you know the problem, the more experience you have with numerical techniques, the better your guess will be. Now, with that N, get a solution and compare it to available data. Is it close enough for all practical purposes? If yes, you are done. More than likely, though, your answer will be no. Your calculated deflections may be some 20% off from analytically determined or measured deflections. You should be able to compute displacement-type quantities to within 2% (assuming your material model is good). Now, double N and obtain another numerical solution. Again, compare it to whatever is available for the variable of most interest to you. Once the numerical and analytical–experimental solutions compare favorably, quit. If they do not, double N again and keep doing this until convergence occurs or you run out of money or computer storage capacity, whichever comes first.

Back to our problem. At each of the $(L/2)/N$ points from $x = 0$ to $x = L/2$ (call these "mesh points"), replace each term of the DE with the finite-difference approximation, as shown. What has happened? We have replaced our original single differential equation with N algebraic equations—and N may eventually go to 100 or

Introduction to Numerical Modeling of Fast, Transient Phenomena 115

FINITE DIFFERENCES

TO DISCRETIZE:
DIVIDE REGION INTO N EQUAL SECTIONS EACH OF SIZE $\frac{(L/2)}{N}$

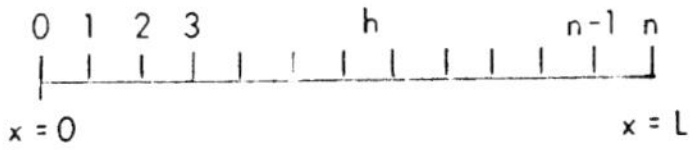

```
   0 1 2 3         h         n-1 n
   |_|_|_|_._|_|_|_|_|_|
   x = 0                     x = L
```

CALL JOINTS "NODAL POINTS" OR "MESH POINTS." LET x, y VALUES AT JOINTS BE x_J, y_J (J = 1, N).

REPLACE EACH TERM OF D.E. WITH FINITE DIFFERENCE APPROXIMATION AT NODES.

EXAMPLE:

AT POINT 1: $EIh^{-2}\left(y_0 - 2y_1 + y_2\right) = \frac{W(x_1)}{2} x_1 (L - x_1)$

AT POINT 2: $EIh^{-2}\left(y_1 - 2y_2 + y_3\right) = \frac{W(x_2)}{2} x_2 (L - x_2)$

- -

AT POINT N-1: $EIh^{-2}\left(y_{N-2} - 2y_{N-1} + y_N\right) = \frac{W(x_{N-1})}{2} x_{N-1} (L - x_{N-1})$

Figure 4.11 Discretization of a continuous beam into *N* sections.

1000 or greater, depending on our accuracy requirements. If we are simulations a well-behaved function, *N* might be quite small. However, if our function is wildly oscillatory, an *N* of 10,000 may not be large enough; that is the penalty we pay in this transformation.

We now need to do the same for our boundary conditions, but here we must resort to some trickery. If we proceed as we did with the PDE, we will find that our finite-difference approximation at the boundary is only *first-order accurate*. This reduces the order of accuracy for the entire problem to first order. We would prefer to maintain second-order accuracy throughout. In order to do this we introduce a fictional point just outside our interval, as shown in Fig. 4.12. Now, with some straightforward algebraic manipulations we eliminate the deflection at the fictional $n + 1$st point and we have our algebraic equations ready to solve. Write them in matrix form (Fig. 4.13). The coefficients of the matrix are known, the components of the load vector are known. Get a matrix equation solver from your system and within fractions of seconds a solution is available for the unknown displacements. Compare these with the exact solution and proceed as above.

FINITE DIFFERENCES

AT BOUNDARIES:

$y(0) = 0 \rightarrow y_0 = 0$

$y'(L/2) = 0$

TO MAINTAIN $O(H^2)$ SCHEME, INTRODUCE FICTIONAL NODAL POINT AT $x = \frac{L}{2} + H$.

WITH THIS N+1st POINT, N BECOMES INTERIOR POINT.

$H^{-2}\left(Y_{N-1} - 2Y_N + Y_{N+1}\right) = F_N$

$0.5H^{-1}\left(Y_{N+1} - Y_{N-1}\right) = y'(L/2) = 0$

$\left.\begin{matrix}\\\\\\\\\end{matrix}\right\}$ ELIMINATE Y_{N+1}

$2H^{-2}\left(Y_{N-1} - Y_N\right) = F_N$

WHERE $F_J = \frac{W(x_J)}{2} \times (L-x_J)$ $(J = 1, N-1)$

Figure 4.12 Boundary treatment to maintain second-order accuracy.

FINITE DIFFERENCES
DISCRETIZED EQUATIONS IN MATRIX FORM

$$\frac{EI}{H^2}\begin{bmatrix} -2 & 1 & 0 & 0 & \cdots & 0 \\ 1 & -2 & 1 & 0 & \cdots & 0 \\ 0 & 1 & -2 & 1 & \cdots & 0 \\ \cdots & \cdots & \cdots & \cdots & \cdots & \cdots \\ 0 & \cdots & \cdots & \cdots & 1 & -1 \end{bmatrix} \begin{Bmatrix} Y_1 \\ \cdot \\ \cdot \\ \cdot \\ Y_N \end{Bmatrix} = \begin{Bmatrix} F_1 \\ \cdot \\ \cdot \\ F_{N-1} \\ 0.5F_N \end{Bmatrix}$$

OR

$[K]\{y\} = \{F\}$

K: N X N SYMMETRIC MATRIX
Y: VECTOR OF UNKNOWN DISPLACEMENTS AT JOINTS
F: LOAD VECTOR

Figure 4.13 Finite-difference equations in matrix form for the beam problem.

Introduction to Numerical Modeling of Fast, Transient Phenomena 117

What did we do? We went from a single differential equation to a system of coupled algebraic equations, which can be compactly written in matrix form using finite differences.

$$[\mathbf{K}]\{y\} = \{F\}. \tag{4.1}$$

b) **Finite elements.**

The finite-element method is sometimes referred to as an *exact* solution to an *approximate* problem. Unlike finite differences, where we manipulate the governing equations of a problem into DE form and then systematically replace derivatives with difference expressions, in finite elements we begin by discretizing our continuum *at the outset*. Thus, our approximation is introduced first—that the continuum with its infinite degrees of freedom can be effectively represented by a discrete system with a finite number of degrees of freedom—and then the equations which result from that approximation are solved exactly.

Consider Fig. 4.14. It endeavors to represent half of a symmetric problem of a short solid cylinder striking a plate with finite dimensions. The figure contains quite a bit of information. To begin with, it states that:

- We have chosen to represent our continuum by a finite system of triangles (60 in the projectile, 280 in the target).

- These triangles will interact with each other *only* at connection points, or nodal points, in the finite-element jargon. If we need information anywhere else in the grid, we will get it by interpolating the nodal point information.

- This finite system of triangular elements and nodal points represents the behavior of the continuum that it replaces to an acceptable degree of accuracy (we hope).

Figure 4.15 elaborates on this a bit and introduces a critical notion in finite elements, the existence of an approximating or interpolating function. Once we solve the system of equations, which comes from approximation, we will have displacements only at our nodal points. Then using the equations of elasticity and plasticity (if we are dealing with deformable solids), from displacements we can get strains. Knowing the constitutive law, from strains we can get stresses and pressures. From stresses and element volumes we can get forces. Without interpolating functions, though, we would have solutions for our unknowns at the nodal points. We want more—we want information within our entire domain. These interpolating functions are usually polynomials. They cannot be selected

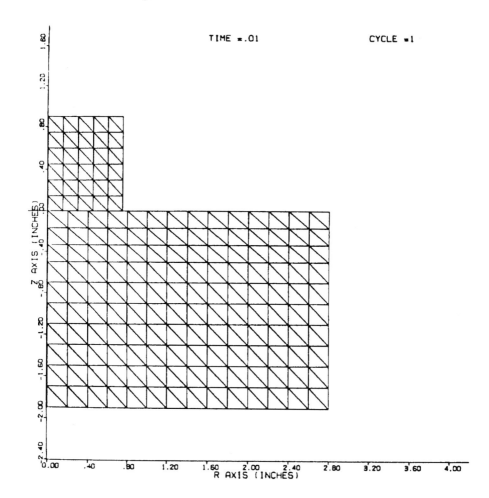

Figure 4.14 Discretization by triangles.

at random, however (Fig. 4.16). They must meet certain requirements to produce physically plausible solutions. Furthermore, in the limit, as more and more elements are used in our mesh and element size tends to zero, the interpolating functions must be able to accommodate *rigid-body motion* and be able to *reach a state of constant strain as the mesh is refined*.

There are numerous books on finite elements. In my opinion, two of the best are the texts by Bathe [1982] and Cook [1995]. At least one of these should be in your library.

Let us again look at a practical problem, this time the response of a beam subjected to an end load P. Before manipulating any equations, we discretize our continuum at the outset (Fig. 4.17) into n elements having nodal displacements of $u_1, u_2, ..., u_n$. An advantage of finite elements is that we can work in a local

FINITE ELEMENT METHOD

DISCRETIZE AT OUTSET

OUTLINE OF PROCEDURE:

- DIVIDE THE CONTINUUM INTO A FINITE NUMBER OF REGIONS CALLED "ELEMENTS"

- THE ELEMENTS INTERACT ONLY AT A DISCRETE NUMBER OF POINTS CALLED "NODES"

- POSTULATE FUNCTIONS TO DEFINE DISPLACEMENTS WITHIN EACH ELEMENT IN TERMS OF NODAL DISPLACEMENTS

- DEFINE THE STATE OF STRAIN FROM DISPLACEMENT FUNCTIONS

- DEFINE THE STATE OF STRESS FROM THE CONSTITUTIVE RELATIONSHIP

- DEFINE ELEMENT NODAL FORCES

- ASSEMBLE GLOBAL ARRAYS

- SOLVE FOR NODAL DISPLACEMENTS USING STANDARD TECHNIQUES

Figure 4.15 Steps in the finite-element discretization process.

FINITE ELEMENT METHOD

CHOOSE APPROXIMATING FUNCTIONS:

- THE NUMBER OF TERMS IN THE CHOSEN SERIES MUST EQUAL THE TOTAL NUMBER OF DEGREES OF FREEDOM OF AN ELEMENT

- THE APPROXIMATING FUNCTION AND CERTAIN OF ITS DERIVATIVES MUST BE CONTINUOUS WITHIN THE ELEMENT. THERE MUST BE COMPATIBILITY BETWEEN ADJACENT ELEMENTS.
  THE APPROXIMATING FUNCTION MUST BE ABLE TO ACCOMODATE:
  - ** *RIGID BODY MOTION*
  - ** *A STATE OF CONSTANT STRAIN IN EACH ELEMENT AS THE MESH IS REFINED*

Figure 4.16 Criteria for interpolating functions.

120 *Introduction to Hydrocodes*

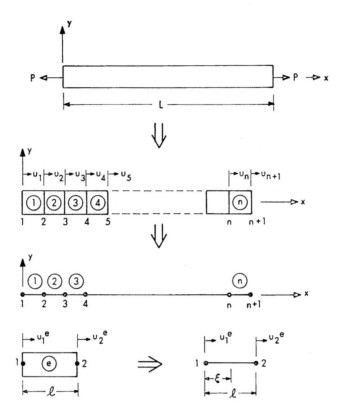

Figure 4.17 Discretization of an end-loaded beam.

coordinate system that simplifies element calculations, then assemble the elements in a global framework. For this particular illustration, it is just as convenient to stay in the global (x,y) system than to go to a local coordinate, as shown at the bottom of Fig. 4.17, but it can be done if so desired.

Now that we have a discrete system, Figs. 4.18 and 4.19 show the procedures to be followed. First, pick an interpolating function. For this problem we will assume that displacements vary linearly in an element. The displacement at any point in an element is then obtained from our nodal displacements, which are the unknowns of the problem.

Having nodal displacements, we can get strains from the strain–displacement relationship. Having strains, we get stresses from the constitutive equation. We can get the element stiffness matrix in a number of ways. The usual approach is through a variational principle. If need be, we now assemble local stiffness matrices and displacement vectors into global arrays, impose boundary and compatibility conditions

Introduction to Numerical Modeling of Fast, Transient Phenomena

FINITE ELEMENT METHOD

EXAMPLE: ASSUME DISPLACEMENT VARIES LINEARLY WITHIN A TYPICAL ELEMENT. CHOOSE:

$$u^e(\zeta) = a_0 + a_1 \zeta$$

TO DETERMINE DISPLACEMENTS AT ANY POINT IN ELEMENT IN TERMS OF NODAL DISPLACEMENTS, EVALUATE ABOVE AT

$\zeta = 0, \ell$

$$u^e(0) = u_1^e = a_0$$

$$u^e(\ell) = u_2^e = a_0 + a_1 \ell$$

IN MATRIX FORM:

$$\begin{Bmatrix} u_1^e \\ u_2^e \end{Bmatrix} = \begin{bmatrix} 1 & 0 \\ 1 & \ell \end{bmatrix} \begin{Bmatrix} a_0 \\ a_1 \end{Bmatrix} \quad \text{OR}$$

$$\{u\}_0^e = [A] \{a\}$$

Figure 4.18 Selection of interpolation function.

FINITE ELEMENT METHOD

FROM STRAIN-DISPLACEMENT RELATIONS:

$$\{\epsilon\} = [B] \{u\}_0^e$$

FROM CONSTITUTIVE EQUATIONS:

$$\{\sigma\} = [D] \{\epsilon\}$$

ELEMENT STIFFNESS MATRIX FROM ENERGY CONSIDERATIONS OR STATICAL EQUIVALENCE OF NODAL FORCES WITH EDGE STRESSES:

$$[K]^e = \int_e [B]^T [D] [B] \, d\,(\text{VOLUME})$$

ASSEMBLE LOCAL STIFFNESS MATRICES AND DISPLACEMENT VECTORS INTO GLOBAL ARRAYS.

IMPOSE BOUNDARY AND COMPATIBILITY CONDITIONS.

Figure 4.19 Incorporating physical properties and applying variational principle.

and, in matrix form, arrive at the expression:

$$[\mathbf{K}]\{u\} = \{F\}. \tag{4.2}$$

Look familiar? What does this tell you? It says very clearly that finite elements and finite differences are merely a means to an end. Both techniques produce a set of algebraic equations which can be attacked with an arsenal of proven methods.

The above was done for static problems. Had we postulated problems involving time-dependence, the resulting equations from either technique would have looked like this:

$$[\mathbf{M}]\{\ddot{u}\} + [\mathbf{C}]\{\dot{u}\} + [\mathbf{K}]\{u\} = \{F\}. \tag{4.3}$$

The matrices $\mathbf{M}$, $\mathbf{C}$ and $\mathbf{K}$ are referred to, respectively, as the mass, damping and stiffness matrices. In many dynamics problems, little is known about damping characteristics so that often the damping matrix $\mathbf{C}$ is not used. When explicit integration schemes are employed to advance solutions in time (this happens in all hydrocodes) and the mass matrix is symmetric (this occurs when mass is lumped at the nodes in place of using a distributed mass for an element), then the global stiffness matrix can be bypassed and the integration performed on an element-by-element basis. This is described in Bathe [1982] and results in a very efficient scheme, which does not require evaluation or updating of the global stiffness matrix. In this case, the equations of motion can be written as:

$$[\mathbf{M}]\{\ddot{u}\} = \{F\}^{\text{External}} - \{F\}^{\text{Internal}}. \tag{4.4}$$

This is the approach used in virtually all Lagrange wave propagation codes.

Remember that wave propagation codes are designed to treat problems with sub-millisecond loading and response times where steep stress gradients or shockwaves are present. This imposes certain restrictions on the numerical approach:

- Since it is inefficient to track shock waves with higher-order polynomials (lots of computing cycles and numerical problems with small gains in accuracy) finite-difference schemes used in wave propagation codes tend to be first- or second-order accurate. Analogously, only linear and constant strain continuum elements are used in the finite-element codes. Just the opposite happens in structures codes. Because the gradients that exist are not as severe and do not move as rapidly with time, higher-order schemes can result in improved accuracy. Structures codes also employ various element types such as beams, plates, shells and other structural configurations in addition to continuum elements.

– Explicit integration algorithms are used exclusively for reasons of accuracy, efficiency and ease of implementation. This will be shown later.

4.3   Lagrangian mesh descriptions.

Lagrangian codes dominate the world of explicit codes. Most structural dynamics codes are Lagrangian. At least half of the existing wave propagation codes are as well. Situations are encountered where deformations are so severe that even with all the features that have been added to Lagrangian codes in the past decade alternatives are required. Euler codes, arbitrary Euler–Lagrange codes (ALE) and meshless (free-Lagrange) methods are outlined in Chapter 6. Hybrid systems have also been used for special applications. These include linked Euler–Lagrange and mappings from Euler to Lagrange and back. With the exception of Chapter 6, the remainder of the book concentrates on Lagrangian meshes.

a)   Mesh characteristics.

We need at the outset a frame of reference in which to solve our equations. In a Lagrange system, the calculation follows the motion of fixed elements of mass. An example often cited is that of a policeman following a single vehicle on a busy highway. His focus is on that one vehicle and he tracks its motion with accuracy. By contrast, in a purely Eulerian system, a fixed grid is established and mass flows through the grid. In our policeman analogy, our traffic cop is now sitting behind a billboard watching cars drive by. He sees the general pattern of traffic flow past some observation point but can say little about the motion of an individual car overtime. In the Lagrange case, we have fine detail and time history data for all the cars being observed. In the Euler case there is an appreciation of the overall flow pattern but at the loss of some detail for individual cars. For some applications, it is convenient to be able to use features of both Euler and Lagrange techniques. Thus, there exists a number of codes which have both Euler and Lagrange processors and elaborate techniques for passing information from one to the other.

Both Euler and Lagrange systems, Figs. 4.20 and 4.21, have their advantages and disadvantages. A major problem of Euler codes is determining material transport. Since material flows through a fixed grid, some procedure must be incorporated in the code to move material to neighboring cells in all the coordinate dimensions. Different schemes

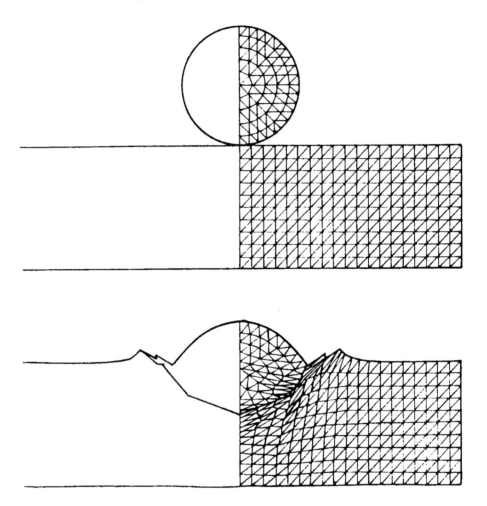

Figure 4.20 Lagrangian mesh.

exist in different codes. None has achieved a position of dominance. It is also necessary to identify the materials in some fashion so that pressures can be calculated in cells carrying more than a single material and stresses and frictional effects determined at interfaces between materials. This is quite difficult for an Euler code to do since it is designed for large flow problems. Hence Euler codes are ideal for large deformation problems but contact is very difficult to determine without adding Lagrangian features.

Lagrange codes, by contrast, have a grid embedded with the material. Because Lagrange codes track the flow of individual masses, the grid deforms together with the material. Hence, Lagrange codes are conceptually more straightforward than Euler codes since there are no transport algorithms to deal with and they require fewer computations

Introduction to Numerical Modeling of Fast, Transient Phenomena 125

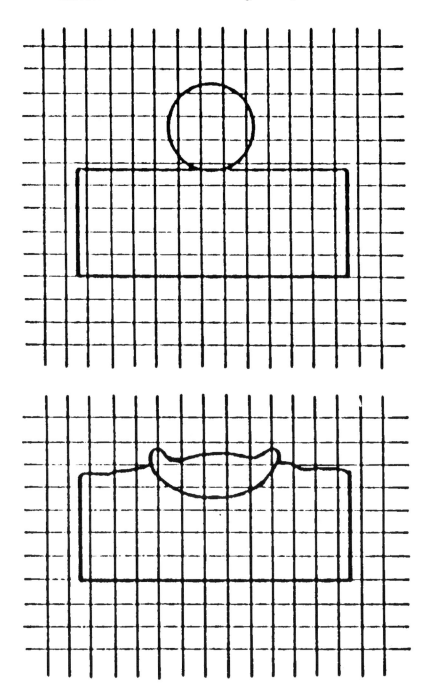

Figure 4.21 Eulerian mesh.

per cycle. Because the grid distorts with the material, time histories are very easily determined with Lagrange calculations, as are material and structural interfaces.

Sounds good so far, does it not? Why does not the whole world use Lagrange methods? The reason is a fundamental law of Mother Nature—there is no free lunch! In order to handle material interactions and contact, Lagrange codes require some complex logic to account for void opening and closing, sliding with or without friction and, in the case of interior failure, creation of new contact surfaces. The more realistic this contact logic (known in the literature as slide lines, sliding interfaces, contact–impact, contact processors), the more accurate the computational results. However, sliding interface or contact logic is computationally intensive. About 80% of the CPU time in a Lagrange calculation is spent in the contact processor. Hence, the more computationally efficient this part of the code, the faster the code, the cheaper the calculation. Because of the importance of the contact algorithm in a Lagrange code, we need to take a closer look at it.

b)   *Contact–impact.*

At any given time step, including the first, *all* nodes in a Lagrangian grid are moved, the displacement being proportional to the velocity assigned to each node. Picture the problem of a moving projectile striking a stationary target. Only projectile nodes have a velocity, so after the motion is applied a portion of the projectile is buried in the target (Fig. 4.22). Clearly, this is a physically unrealistic situation. Problems also occur when large relative displacements occur, such as in hypervelocity impact calculations, where fluid-flow situations exist and slippage of one grid over another must be taken into account. This aspect of material motion is accounted for in Lagrange codes by the contact processor, a currently popular buzzword. Initially, the subroutines accounting for sliding and inter-penetration were referred to as slide lines or sliding surfaces.

Contact processors have been a part of Lagrange hydrocodes since the appearance of HEMP, developed by Mark Wilkins and his associates at the Lawrence Livermore Laboratories, in 1964. The sliding interface algorithm used in HEMP is described in Wilkins [1999]. The algorithms that followed tended to follow the work of Wilkins since by and large hydrocodes are developed by evolution rather than revolution. Equivalent formulations for finite elements were first put forward by Johnson [1978] for EPIC and Hallquist [1988] for DYNA. Valuable publications on the subject are Hallquist [1976, 1978], Goudreau & Hallquist [1982], Hallquist, Goudreau & Benson [1985], Benson [1992] and Wilkins [1999]. A description of the computer codes mentioned in this section can be found in Zukas [1990].

For a long period following the introduction of the earliest hydrocodes (HEMP, EPIC, DYNA on the Lagrange side; HELP and HULL on the Euler side) in the mid-1960s

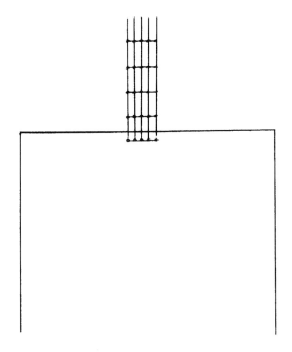

Figure 4.22 Inter-penetration at the start of a calculation.

through mid-1970s, calculations involving contact were performed by a small group interested in space applications (impacts in the area of 6 km/s), military problems (impact velocities in the range of 1.0–1.5 km/s) or basic research in the behavior of materials at high strain rates. The slide-line logic in hydrocodes fell into two categories: (1) a node, which penetrated a sliding surface was put back on that surface on a path that kept it on its original trajectory. Then linear and angular momentum conservation was applied and the velocities of the nodes involved in the penetration adjusted; or (2) a node penetrating a sliding surface had a restoring force assigned to it and was gradually returned to the appropriate surface. The latter approach worked well for low-velocity problems such as structural impact but the first was better suited for high-speed impacts. Most of the procedures for treating pathological problems were developed on an *ad hoc* basis over a number of years. A good discussion of these is given by Breidenbach [1997]. Difficulties with contact problems are being documented and available in the Internet websites for many codes [e.g., ABAQUS, 2001]. The most elaborate contact processor available in production codes is incorporated in the DYSMAS-L code [Poth *et al.*, 1981, 1983, 1985].

For the past two decades, with the advent of technology, the same problems of detecting and modeling intrusions in rigid bodies has become of interest in problems of robotics, computer graphics as welt as structural impact problems in civilian applications such as safe design of containment structures and strengthening structures against terrorist attacks using explosives or projectiles. See, as representative (but hardly comprehensive) examples, Taylor & Papadopoulos [1993], Papadopoulos & Taylor [1993], Mottershead [1993], Konter [2000], Ponthot & Graillet [1998], Baraff [1989; 1994; 1996]. Much of this research has emanated from universities and research centers focusing primarily on structural dynamics. This group, unfamiliar with the earlier work with hydrocodes, has relied primarily on various Lagrange multiplier techniques and penalty methods. Lagrange multiplier methods ensure non-penetration by determining penetration correction forces and displacements. The method does not need any additional parameters and accommodate models of sliding friction. Penalty methods apply penalty forces to penetrating nodes and let time integration correct penetrations. Penalty methods work well in situations where energy release occurs over a fairly long time-scale (milliseconds to seconds) as, for example, in vehicle crashes. Penetrations need not be corrected exactly at each time step and contact forces are not as important as global parameters. In contrast, for high-speed dynamics problems such as impact and penetration, the relevant physics occurs in a small zone. Thus, stresses and wave propagation need to be determined across contact zones with great accuracy. Lagrange multiplier methods are better suited to these problems [Diekmann et al., 2000]. Very formal and sometimes complex mathematical procedures are derived to solve the contact–impact problem. These are generally found to be computationally expensive and recourse is made to linearization to run problems at finite cost.

No slouches either, the hydrocode community has actively been developing and refining contact algorithms as well. In many codes, nodes or surfaces involved in contact need no longer be specified by users, a tedious procedure and sometimes costly if the contact area is not specified to be large enough and a problem with a runtime of 20–100 h must be aborted and restarted. Symmetric search techniques on contact surfaces have been developed, so that, in many codes, the designation of master and slave surfaces is arbitrary. Contact algorithms for parallel processing are being actively researched and developed.

It has now become clear that both communities are using the same techniques, although they arrived at them from different starting points. The put-back logic of hydrocodes is the equivalent of a Lagrange multiplier technique. The assignment of a restoring force to a node penetrating a contact surface is the equivalent to the penalty method. Both techniques include various *ad hoc* treatments for pathological intrusions, rules of thumb for detecting contact, restoration of penetrated nodes, etc. Because of their failure to research the literature, the group working the structural impact–robotics–graphics problems has spent 40 years developing, testing and refining techniques that have been available since the 1960s. Even more remarkable is that many of these researchers are based at universities. It used to be a rule that a graduate student, before embarking on a research project, had to make a thorough review of the literature to avoid duplicating work already done. Apparently, this is either no longer enforced, does not apply to professors or both. This is not a complete loss. Thanks to current research efforts much of the algorithmic work in contact–impact is now on sounder mathematical footing and this is a good thing. Still, from the standpoint of practicality, much time and money has been wasted rediscovering techniques that have been available since at least 1964!

As Konter [2000] points out, contact is not described directly by finite-element equations. Hence, procedures must be specified. In order to keep the running time manageable, some method for narrowing the processing of penetration and sliding must be limited to only the nodes involved. Contact events occur at non-predictable points in time and at non-predictable places on various surfaces. Thus, a search of contact events must be made on all surfaces at every time step. Examples of efficient searches are given by Segletes & Zukas [1993] and Diekmann *et al.* [2000], among others.

Contact surfaces (or sliding interfaces) are required whenever contact or separation of materials in a calculation can be expected. Typical examples include situations involving large density gradients such as the interactions of gases and fluids with solid walls, situations involving penetration of one body by another, either rigid or deformable contact between colliding bodies, regions where large shear distortions occur and internal fracture resulting in material separation and perhaps recompression.

At the outset, it is necessary to designate the material surfaces that are expected to experience contact during the calculation (Fig. 4.23). This is done by specifying master and slave nodes (or surfaces). In the earliest codes this function was performed by the user and was part of the code's input package. Many guidelines existed as to which surfaces should be declared slave or master. Only the master surface was checked for intrusion. Today it is the custom to check intrusion of both slave and master surfaces, so that the designations have become irrelevant. In addition, the pre-processors to today's hydrocodes generate much of the interface information. The ultimate example of this is the *ZeuS* code where all free surfaces and all material interfaces are designated as contact surfaces. The user may eliminate some of these from consideration depending on the physics of the problem but is never required to specify a list of nodes. Since this information is already generated by the pre-processor, it is pointless, as well as a source of error, to require the user to repeat what the computer has already done.

Once contact surfaces are specified, at each computational cycle the velocities and the accelerations of nodes involved in contact are decomposed into normal and tangential components. In the *normal* direction, motion of the nodes is continuous

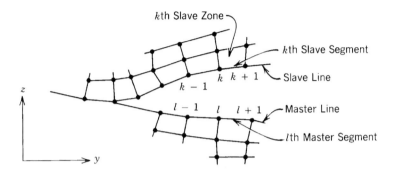

Figure 4.23 Master and slave nodes.

during contact, independent when separated. In the *tangential* direction, the motion of nodes is independent when materials are separated or if the interface is frictionless. The motion is restricted if there is contact or friction is present. Sliding friction can be incorporated in hydrocodes. When this is done it is usually by way of a Coulomb model. Some codes also incorporate *tied* sliding displacement compatibility is maintained at abrupt grid changes. The requirement of no separation or sliding may be maintained throughout the calculation or removed under conditions specified by the user.

The computational procedure for contact in Lagrangian systems consists of the following:

(1) Identify a series of nodes that make up the master surface.

(2) Identify a series of nodes that make up the slave surface.

(3) For each increment in time, apply the equations of motion to both master and slave nodes.

(4) Check for interference between the designated contact surfaces (slide lines) by

- defining a search region;
- checking each slave node for intrusion into the master surface. If penetration of the contact has occurred, do something. The two most popular choices are to move the slave node to the master surface in a direction perpendicular to the master surface (a.k.a. Lagrange multiplier method) or insert a linear spring type restoring force to gradually bring the offending node back to the master surface (a.k.a. the penalty method). A third choice is to reduce the time step to just that required for a potentially intruding node to just land on

the master surface. This, however, is computationally intensive and is no longer in vogue.

(5) Once the protruding slave node is back on the master surface, linear and angular momentum conservation is invoked, frictional forces, if any, are applied and voids are opened (if tensile forces are present).

(6) The above steps are repeated for each computational time step.

The astute reader will have noted a few difficulties with this deceptively simple procedure, the heart of any Lagrangian code which includes contact–impact capability. Here are but a few.

Restoration of a penetrating slave node to its corresponding master surface in a direction perpendicular to that surface only works if the choice of "corresponding" is obvious. For situations involving corners or craters, the choice may not be obvious. Then, criteria need be specified as to which master surface is the proper one. This requires storing the node's current as well as previous position so that its path may be determined. Repositioning a node to an improper master surface can lead to artificially large distortions once element areas and volumes are computed and lead to large changes in energy in a single computational cycle. Many examples are discussed by Breidenbach et al. [1997]. Specification of too large a restoring force in the penalty approach may also cause large energy jumps. Specifying one too small will cause the node to continue to drift end artificially distort the element to which it is attached.

Once a node is restored to its corresponding master surface, conservation of momentum is applied and velocities of the slave and the segment's master nodes are adjusted. So far, so good. What happens if the next segment in line is penetrated as well? The same procedure is applied and its nodal velocities are adjusted. However, since the segments share a common node, the velocities of the previous segment are no longer in balance. These problems can sometimes be corrected by a double-pass approach, where the designation (master and slave) of the contact nodes is reversed and the procedure outlined above is repeated.

Related to this is the problem of several nodes penetrating a single master surface. The contact logic in early Lagrange code restored these sequentially. The first intruding node found by the search algorithm was moved to the master surface and momentum balance and contact conditions enforced. Then the second, third, etc. nodes were treated accordingly. With each node treated, the velocities of other nodes on the surface were altered, causing an interface phenomenon known as "rattling." The *ZeuS* algorithm, among others, employs a simultaneous treatment for all slave nodes, which intrude a given master surface. As a result, surface rattling is greatly diminished (see Fig. 4.24).

132 *Introduction to Hydrocodes*

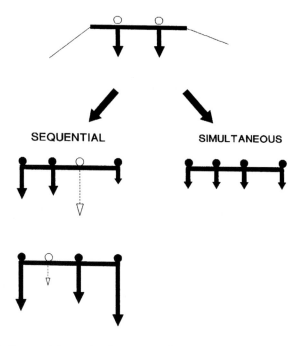

Figure 4.24 Sequential vs. simultaneous adjustment of penetrating slave nodes to minimize rattling.

These are only a few examples of problems which can arise in dealing with contact. In practice, an algorithm is written in following steps (1)–(6) above by the code developer. The code users then encounter the various pathological problems that arise. Hence, unless the code is written by the group that uses it, a close relationship with the code developer is necessary to make modifications as they are needed. Typically, the contact algorithm in a Lagrange code consists of but a few of the hundreds of subroutines in the code, with the exception of DYSMAS-L. Yet it easily takes up at least 80% of the run time for any problem and accounts for nearly 100% of the hair loss among code users. Its effects are usually felt further along in a computation. The calculation will stop citing excessive energy growth or unrealistic pressures from the EOS or a reduced time step which makes the calculation economically unfeasible (Figs. 4.25 and 4.26).

A novel development in contact surface treatment is the Pinball Algorithm developed by Neal and Belytschko. (See Neal [1989] and Belytschko & Neal [1989] for details and examples. Calculations of deformable projectile ricochet from deformable plates with this algorithm have been performed by Zukas & Gaskill [1996].) The main concept of the Pinball Algorithm is to enforce the impenetrability condition and contact

Introduction to Numerical Modeling of Fast, Transient Phenomena   133

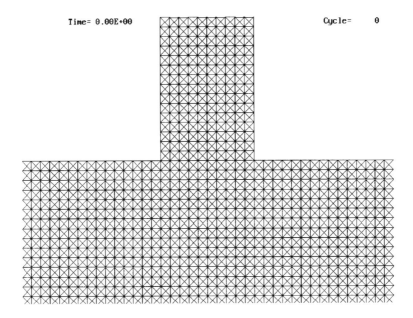

Figure 4.25 Initial grid for projectile penetration calculation [Segletes and Zukas, 1996].

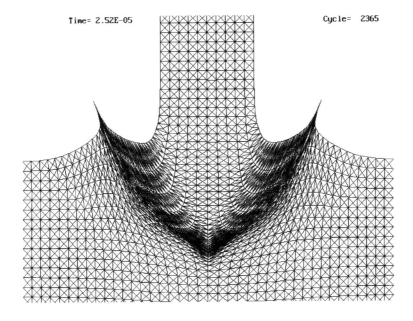

Figure 4.26 Extreme mesh distortion in calculation of projectile penetration [Segletes and Zukas, 1996].

conditions on a set of spheres, or pinballs, which are embedded in the finite elements. In the first cycle, the radius needed to create a sphere of volume equal to the element volume is computed for each element. For elastic–plastic problems most of the element deformations can be considered neatly incompressible. Therefore, the element volume, and also the radii of the pinballs, will change little over the course of a simulation. For this reason, the radii are calculated only once. For penetration detection on subsequent cycles, the distance between the centers of each slave pinball and each master pinball are calculated and compared with the sum of the radii of the two elements. If the distance is less than that sum, penetration has occurred and a corrective force (note the relationship to the penalty method at this stage) is applied.

The Pinball Algorithm greatly reduces the time required in the search for penetration over conventional search algorithms. Inter-penetrability becomes a simple check of distance between two pinballs and, since it involves almost no recursive calculations it lends itself readily to vectorization. As with any new algorithm, some trial and error calculations are needed to uncover its eccentricities. Zukas & Gaskill [1996] found that the Pinball Algorithm works very well for impact calculations that are finely meshed in the vicinity of contact. Where element size was too large, element aspect ratios were greater than one, or in low-speed impacts where time of contact is a crucial parameter; then the Pinball Algorithm can behave in a less than ideal manner. Because in 3D calculations penetration by hexagonal elements is treated as penetration by spheres, penetration detection may be delayed at element corners and be premature at element faces. The appeal of the procedure is in the time savings for high-speed impact problems with appropriate meshes.

c)   ***Large distortions.***

A second problem with Lagrange codes is large distortions. As the grid distorts with the material, element size tends to zero. We will see in the next section that the time step is based on the size of the smallest element in the grid. Hence, as element size tends to zero, the time step tends to zero. Lots of computational cycles are burned up but little progress is made. It can get worse, though, depending on the type of element used. Quadrilaterals and hexahedrals can invert under large distortions. This causes a condition where negative volumes, and therefore negative masses, are computed and the calculation becomes invalid. Triangular elements will not invert since this would require them to go through a state of infinite pressure. However, gross distortion of triangles can cause the grid to become artificially stiff locally and lock up along 45-degree lines. Again, the calculation is rendered useless.

Introduction to Numerical Modeling of Fast, Transient Phenomena   135

Two techniques are available to overcome these problems. The most general is *rezoning*. When the time step becomes too small (by some user-defined criterion), a restart file is written. Then, a new, undistorted grid is overlaid on the distorted one and mass, momentum, energy and the constitutive equation are conserved. This allows the calculation to continue but at some cost. The history of several elements distorted elements may be mapped into a single, larger, undistorted element, thus causing diffusion. In fact, if rezoning were done every cycle we would have a Eulerian code. Modern Euler codes use a two-step procedure, the first being a Lagrange computational step followed by material transport, which effectively restores the Euler grid to its undistorted (rezoned) form. Repeated rezoning causes smoothing and makes Lagrange calculations look very much like Eulerian ones. Some codes like DYNA have automatic rezoners. However, considerable human intervention is still required to be sure that the rezones are physically correct. An example of rezoning is shown in Fig. 4.27.

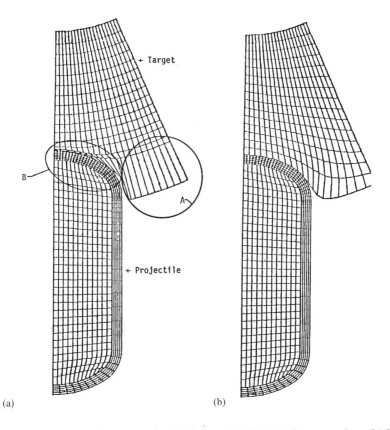

Figure 4.27 An example of a rezoned calculation. (a) Mesh before rezoning. (b) Mesh after rezoning.

136   *Introduction to Hydrocodes*

A second approach works well when distortions are highly localized, such as in rod and jet penetration problems. It is called "erosion" but has nothing to do with the metallurgical description of a particular mode of material failure. Instead, it is a bookkeeping technique that allows dynamic redefinition of sliding interfaces. As elements on an interface fail, the contact surfaces are recomputed on the fly and the distorted elements removed from the calculation. This conserves mass and momentum exactly, but energy only approximately since the internal energy of failed elements is no longer tracked. When it first appeared in EPIC in the U.S. and DYSMAS in Germany the technique was received with some criticism, but has now been incorporated in the AUTODYN, DYNA, EPIC, DYSMAS, PRONTO, *ZeuS* and other codes under names such as "eroding slidelines", "advected elements" (as in "advected" out of the calculation) and other fancy descriptors. An example of the working of the erosion method is shown in Figs. 4.28 and 4.29. The calculation is the same as that in Figs. 4.25 and 4.26 above but with erosion permitted.

An example of the close agreement that can be achieved between experimental results and calculations employing the eroding slide-line technique is shown in Figs. 4.30 and 4.31. These depict the various stages for penetration of a finite-thickness armor steel plate by a steel projectile with length-to-diameter ratio of 10 and mass of 65 g. The striking velocity for the normal impact case was 1103 m/s. The experimental data, reported by Lambert [1978], shows a residual projectile velocity of 690 m/s and a

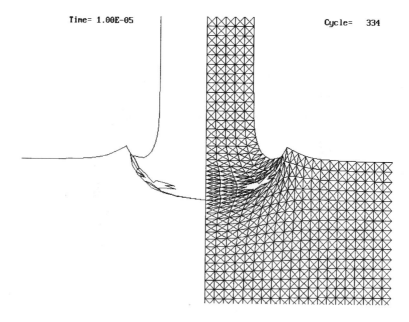

Figure 4.28 Penetration calculation with eroding slide lines at 10 μs after impact.

Introduction to Numerical Modeling of Fast, Transient Phenomena   137

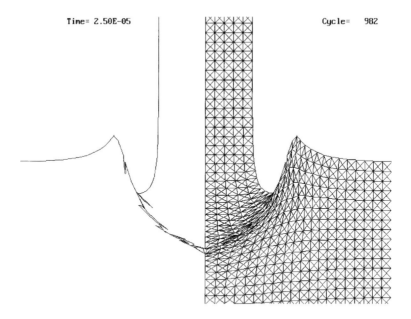

Figure 4.29 Penetration calculation with eroding slide lines at 25 μs after impact.

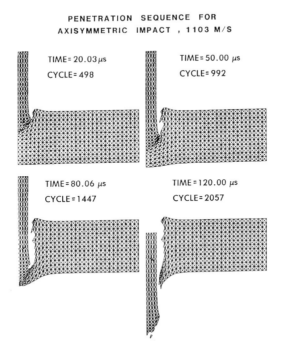

Figure 4.30 Normal impact penetration calculation.

138 *Introduction to Hydrocodes*

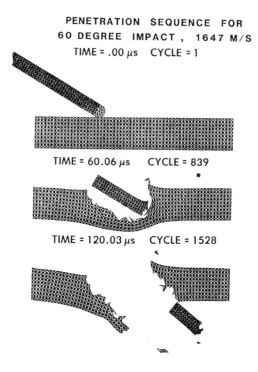

Figure 4.31 Oblique impact calculation.

residual mass of 32.7 g. The calculations with EPIC indicated a residual velocity of 709 m/s and a residual mass of 32.1 g. For the 60-degree impact case, the computed vs. measured residual velocities were 1202 vs. 1145 m/s while residual mass was computed to be 22.9 vs. 16.8 g determined experimentally.

4.4   Artificial viscosity.

Artificial viscosity is included in Euler and Lagrange codes for two reasons:

(a) allow a code with a continuum formulation to handle shock waves, which, mathematically, are discontinuities;

(b) provide grid stabilization for quadrilateral and hexahedral elements which use one-point (reduced) evaluation of the element constitutive model.

The mathematical basis for all structural dynamics and hydrocodes is based on the assumption that we are dealing with a continuum. This, in turn, precludes the presence of nathematical discontinuities. Our physical problems, aterial failure, etc., do indicate the presence of shock e materials we work with is not enough to affect them. uld arise. How does one measure the viscosity of a solid? rematical formulations which allow for internal floating f continuous flow. Thus we could accommodate shock tting. However, incorporating internal boundaries as part done in 1D calculations, is a non-trivial exercise in two plicate the matter we would need to know when shock n and account for their appearance and disappearance. t smart yet. Finally, such shock tracking logic makes heavy resources. While calculations running 100 h or more have o not allow computations to support experimental programs. ootnotes to the literature. ssuming that the term means something for solids and, more e measured at finite cost, is very small for most materials. This ocks and their accurate resolution would require lots and lots of ne step in a calculation is proportional to the size of the mesh, this ccumulation of computational cycles with minuscule advancements

The solution to this dilemma was discovered by von Neumann & Richtmyer [1950] who introduced the concept of an artificial viscosity, added to the pressure, that had the effect of smearing the shock wave over several mesh widths, this converting it from a discontinuity to a steep stress gradient. Steep stress gradients, we can handle. They are not pleasant, but we can treat them. The presence of artificial viscosity does distort the solution but it has two lovely properties:

(a) The solution is affected only at the shock front. As we move away from the front the viscosity no longer affects the calculation.

(b) The accuracy of the calculation is preserved.

The exact form of the viscosity function is somewhat arbitrary. Many variants have been proposed in the literature. The most common form found in codes today is a two-term form shown below. The linear term is used to suppress spurious oscillations behind the wave front. The quadratic term spreads the wave front over several cells in the direction of propagation and lowers the peak amplitude. Excessive application of viscosity damps out the solution.

Here, the cs are user-defined constants, with the most commonly used values indicated. The unsubscripted

$$o = c_1 \rho c \Delta x \left|\tfrac{d\dot{x}}{dx}\right| + c_2^2 \rho (\Delta x)^2 \left|\tfrac{d\dot{x}}{dx}\right|\left|\tfrac{d\dot{x}}{dx}\right|. \tag{4.5}$$

$$0.05 \leq c_1 \leq 0.5. \tag{4.6}$$

$$c_2 \approx 2. \tag{4.7}$$

c is the sound speed of the material. The term Δx represents the mesh width while ρ represents density. Extension of the above formulation to two and three dimensions has been derived by Wilkins [1980]. A very lucid work on the artificial viscosity technique is the report by Noh [1976]. Both references should be in your personal library.

The effect is very much mesh size dependent. Calculations with very fine resolution may not need it, or calculations where extreme gradients do not exist. Use of excessive viscosity in coarsely gridded calculations, especially those involving energetic materials, will cause disturbances to propagate at high, non-physical rates and can lead to instabilities. Experience is the key to success.

Figures 4.32–4.35 show the effects of varying the linear coefficient for an elastic impact of a steel rod onto a rigid surface at 10 ft/s. Excessive use of artificial viscosity can overdamp a solution and suppress meaningful information.

In addition to the von Neumann–Richtmyer viscosity, another type, referred to as hourglass or tensor viscosity, is used in computer programs with under-integrated elements. For example, a four-node quadrilateral element would be integrated (have the state of its material evaluated at each cycle) using a 2×2 Gauss quadrature rule. Evaluating the constitutive model at four points in each element at each cycle is an expensive proposition. To save time and money, it is possible to evaluate the element's stress state at one point in the element—the center. This makes for economical calculations but results in certain types of motions in two and three dimensions that are not resisted by internal forces. An example is the velocity field shown in Fig. 4.36, which leads to an hourglass-type instability not resisted by either internal element forces or the von Neumann–Richtmyer-type viscosity. To retain the advantages and economics of one-point integration but counter destabilizing motions, viscosity expressions are introduced to stabilize the mesh. Not one method has achieved universal acceptance as yet. Various schemes have been proposed by Wilkins [1980], Belytschko et al. [1984], Belytschko & Lin [1985], Jaquotte & Oden [1984], Flanagan & Belytshko [1981], Kosloff & Frazier [1978], Belytshko & Liu [1983, 1986], Schulz [1985], Schulz & Heimdahl [1986] and Verhegghe & Powell [1986], among others.

Introduction to Numerical Modeling of Fast, Transient Phenomena 141

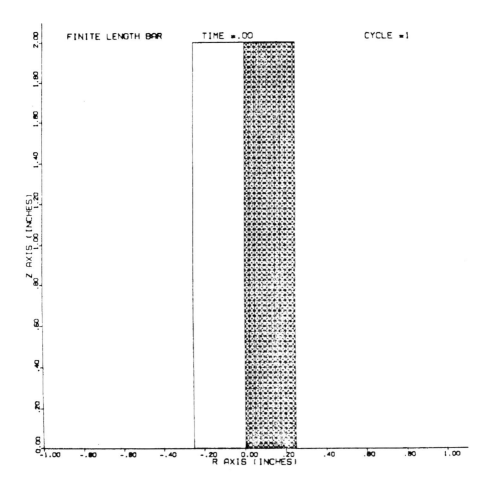

Figure 4.32 Initial geometry.

4.5 Time integration.

Insofar as dynamics problems are concerned, the procedures for advancing solutions in time are divided into two categories:

Implicit methods.

Explicit methods.

Look at Fig. 4.37. This outlines the explicit procedures used in all production wave propagation codes. It is also sometimes referred to as the *central difference method*. The key to differentiating between explicit and implicit methods is in step 2. Everything we

142 *Introduction to Hydrocodes*

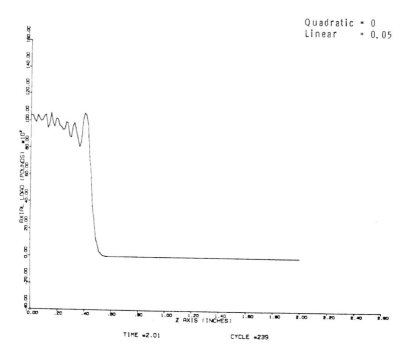

Figure 4.33 Axial force vs. distance for a linear coefficient of 0.05.

want to know at the next time step—velocities and displacements—is on the left-hand side of the equal (=) sign. The quantities that determine these are on the right-hand side *and all are known since they were determined at the previous time step!* That makes for a very simple scheme. We start the problem by determining velocities from initial conditions. From the velocities follow displacements. Having these we get strain rates, strains, stresses, pressures, nodal forces. Now we advance our solution one time step (ΔT). The new velocities and displacements are determined from the old ones and we continue as above, extending our solution as far as desired.

Having learned that there is no free lunch, you now ask "What is the catch?" Good question. Explicit schemes are only *conditionally stable*. Specifically, the time step Δt must satisfy the Courant condition

$$\Delta t \leq 2/\omega, \tag{4.8}$$

where ω is the highest natural frequency of the mesh. In practice it is a royal nuisance to compute natural frequencies, except for the lowest few, so the time step is obtained from

$$\Delta t = \frac{kl}{c}, \tag{4.9}$$

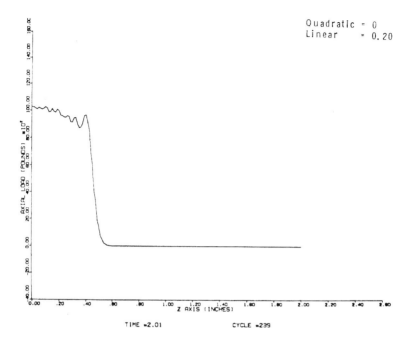

Figure 4.34 Axial force vs. distance for a linear coefficient of 0.20.

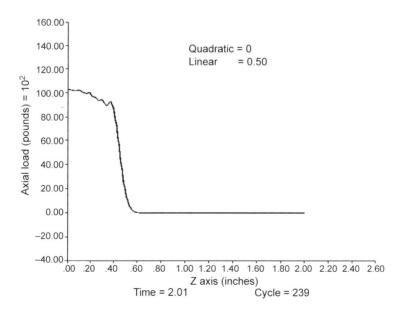

Figure 4.35 Axial force vs. distance for a linear coefficient of 0.50.

144 *Introduction to Hydrocodes*

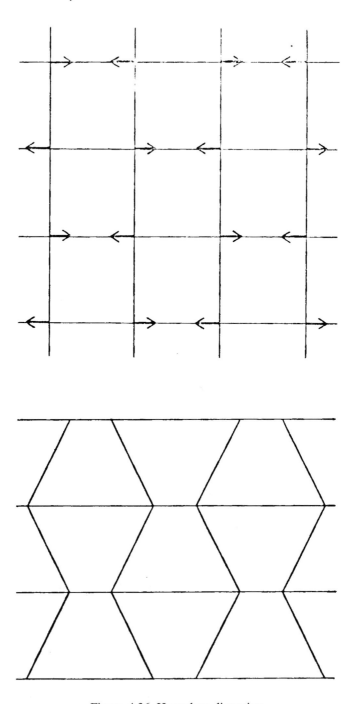

Figure 4.36 Hourglass distortion.

Introduction to Numerical Modeling of Fast, Transient Phenomena 145

NUMERICAL INTEGRATION

FLOW CHART FOR EXPLICIT INTEGRATION SCHEME

1. SET INITIAL CONDITIONS ($\tau = \tau_0$)
   - VELOCITIES $[\dot{u}(\tau)]$
   - DISPLACEMENTS $[u(\tau)]$

2. FIND $\dot{u}(\tau + \Delta\tau/2)$, $u(\tau + \Delta\tau)$ FROM

   $u(\tau + \Delta\tau) = u(\tau) + \Delta\tau \left[\dot{u}(\tau + \Delta\tau/2)\right]$

   $\dot{u}(\tau + \Delta\tau/2) = \dot{u}(\tau - \Delta\tau/2) + \Delta\tau \left[\ddot{u}(\tau)\right]$

3. FIND INTERNAL NODAL FORCES BY LOOPING THROUGH ALL ELEMENTS IN MESH
   - COMPUTE STRAIN RATES AND STRAINS
   - COMPUTE STRESSES FROM CONSTITUTIVE LAW
   - FIND NODAL FORCES FROM ELEMENT STRESSES.

4. COMPUTE EXTERNAL LOAD NODAL FORCES.

5. FIND $\ddot{u}(\tau + \Delta\tau)$ FROM EQUATIONS OF MOTION.

6. INCREMENT FORWARD IN TIME ($\tau = \tau + \Delta\tau$).

7. GO TO 2.

Figure 4.37 Flowchart for explicit integration scheme.

where l is the smallest mesh dimension; c, is the sound speed and k, a stability fraction typically on the order of 0.6–0.9. The stability fraction is there because the above equation was derived from a linear analysis. Throwing in a k in theory (more fervent hope than theory, actually), makes it applicable to nonlinear problems as well. In practice, stability fractions of between 0.6 and 0.8 give good results in both Euler and Lagrange calculations for a wide class of problems. If it becomes necessary to drive the stability fraction to a very low value of 0.4 or less—beware. Very likely there is something wrong with your problem setup. Re-examine both the mesh and the material description very carefully.

One of the best references on numerical techniques for transient problems (and one that has the derivation of stability criteria for many numerical integration schemes) is the book by Belytschko & Hughes [1983].

One way to overcome the problem of using a single time step for the entire grid is to divide the grid into partitions each with its own time step. Let us say that the most severe loading occurs in partition 1. This is where the smallest elements will be with the smallest Δt. Say partition 2 has elements which result in a time step of $3\Delta t$, partition 3 has a time step of 5Δt while partition 4, rather far removed from the action, has rather large elements requiring a time step of 10Δt. When the calculation begins the initial conditions are applied and the entire grid updated. Subsequently, partition one will be integrated two more times with a time step of Δt while the other partitions are ignored. When the clock reaches 3Δt partitions 1 and 2 will be updated. At time equal to 5Δt partitions, 1, 2 and 3 are updated while at 10Δt all four partitions are updated. Thus, partition 1, which sees the most violent distortions, is updated at every time step but partitions 2, 3 and 4 are updated only 1/3, 1/5 and 1/10 as often, respectively. This type of scheme can reduce computing time by some 20–40% and even more. The degree of reduction depends on both the computational grid and on the cleverness of the analyst in selecting the partitions. If the boundary between partition 1 and partition 2 is poorly chosen, much numerical garbage can be generated which can invalidate the calculation. If slide lines overlap the partitions, special provisions must be made to assure that too large a time step is not used for nodes undergoing contact.

Is it possible to have integration schemes that are unconditionally stable and allow for a large time step in the calculation? Yes. Most implicit schemes are unconditionally stable. What is the cost of unconditional stability? Complexity!

The three most widely used implicit schemes are the Newmark β, the Wilson θ and the Houbolt methods. Newmark's scheme is shown in Fig. 4.38. Look carefully at the right-hand side of the equations. In order to obtain velocities and displacements at our advanced time $t + \Delta t$ we must also know the acceleration at the advanced time step, which, of course is an unknown. To get out of this predicament, it is necessary to combine the two equations in Fig. 4.21 with the equations of motion. Now, instead of having a little bootstrapping method with just two equations, we have to solve a systems of simultaneous equations for the displacement at $t + \Delta t$, and generally they are nonlinear! First, then, we must linearize them in some fashion, then solve for the displacement. From such schemes we obtain a much larger Δt than from explicit schemes but at the price of having to solve a system of simultaneous equations at each cycle.

Suppose we are trying to solve a problem in structural dynamics. Let us say a shipping container full of hazardous materials falls from the back of a truck or an airplane and drops on a concrete pad some 5–10 ft away. There will naturally be some crushing of the container and the natural questions are *how much crushing and will the contents leak endangering those nearby and, perhaps, the environment.* Here, loading and response times will be quite long—somewhere between milliseconds and seconds. Details of the wave propagation generated on impact are not needed since it will be the cumulative action of multiple wave reverberations, which will cause damage, not the passage of the first wave. We are interested in the total deflection, which will occur at several seconds after impact. To do this with an explicit algorithm with its time step limited to about a

Introduction to Numerical Modeling of Fast, Transient Phenomena   147

NUMERICAL INTEGRATION

IMPLICIT: DISPLACEMENTS AT $t+\Delta t$ CANNOT BE FOUND WITHOUT
KNOWLEDGE OF ACCELRATIONS AT THE SAME TIME.

NEWMARK-β METHOD:

$$\dot{u}(t+\Delta t) = \dot{u}(t) + \Delta t \left[(1-\gamma)\ddot{u}(t) + \gamma \ddot{u}(t+\Delta t) \right]$$

$$u(t+\Delta t) = u(t) + \Delta t \dot{u}(t) + \Delta t^2 \left[(\tfrac{1}{2}-\beta)\ddot{u}(t) + \ddot{u}(t+\Delta t) \right]$$

PROCEDURE:

   COMBINE ABOVE WITH EQUATIONS OF MOTION.

   OBTAIN A SET OF SIMULTANEOUS EQUATIONS IN $u(t+\Delta t)$.

   SOLVE IF LINEAR.

   IF NONLINEAR, LINEARIZE IN SOME FASHION, THEN SOLVE
   SYSTEM OF EQUATIONS FOR $u(t+\Delta t)$.

Figure 4.38 The Newmark implicit numerical integration scheme.

microsecond or less to keep the scheme stable would require thousands of time steps and get us to the point where we would need to seriously worry about roundoff errors swamping our solution. An implicit scheme, by contrast, with a time step on the order of milliseconds could compute the solution within a few hundred cycles.

What if we are dealing with a hypervelocity impact problem where a projectile strikes a thin plate at very high velocity, say 6 km/s, and produces a debris cloud. Which is the preferred integration method? Here, we need to track wave motion explicitly since it causes failure in both bumper and projectile with just a few wave reverberations. Explicit schemes do this naturally. An implicit scheme could do this problem also. It would allow a much larger time step, but if we used it we would wash out significant details of the problem. For the sake of accuracy, our time step for the implicit scheme would have to be of the same order of magnitude as the explicit scheme. However, at each cycle we would be performing two calculations to update the time in an explicit scheme whereas about 150 calculations would be required for an implicit scheme (remember—system of simultaneous equations). In terms of cost, implicit schemes are not competitive for drain rate problems.

The determining factors in the choice of a time integration scheme are cost and accuracy. As a rule of thumb explicit schemes are more cost-effective for wave

148 *Introduction to Hydrocodes*

propagation problems whereas implicit schemes are better for structural dynamics problems.

Until now, the above rule has been the conventional wisdom. Lately, many computational results have appeared using explicit schemes, which would normally have been done with implicit schemes. The attraction of explicit schemes is their simplicity. In addition, it has been found that roundoff error has not been a factor, even though a large number of cycles are required to reach a solution. Problems formerly addressed with structures codes such as ADINA, MARC, NASTRAN, NIKE, MAGNA and others are now routinely being performed by explicit codes such as DYNA, PRONTO, DYSMAS, AUTODYN and others. These include the shipping container problem mentioned above. For examples, see Brebbia & Sanchez-Galvez [1994], Bulson [1992, 1994], Aliabadi & Brebbia [1993], Aliabadi & Alessandri [1995], Chen & Luk [1993], Shin [1995], Shin & Zukas [1996], and Rajapakse & Vinson [1995].

4.6 Constitutive models.

Explosions result from the sudden release of energy. The energy may come from an explosive, wheat flour dust in a grain elevator, pressurized steam in a boiler or an uncontrolled nuclear transformation. The accumulated energy is dissipated in various ways such as in blast waves, the propulsion of missiles or by thermal or ionizing radiation [Kinney & Graham, 1985]. As the blast wave travels away from its source the overpressure (pressure in excess of atmospheric pressure) at the front steadily decreases and the pressure behind the front falls off in a regular manner. After a short time, the pressure behind the front drops below atmospheric so that an underpressure rather than overpressure exists. During this negative (rarefaction or suction) phase, a partial vacuum is produced. Air is sucked in instead of being pushed away. At the end of negative phase, which is somewhat longer in duration than the positive phase, the pressure returns to ambient. Underpressures usually have a magnitude of less than 28 kPa whereas overpressure can exceed 2 MPa.

Structures suffer damage from air blast when overpressures exceed 3.5 kPa. The distance to which the overpressure level extends depends on the energy yield of the explosion. The blast loading is seen as a lateral dynamic pressure, applied rapidly, lasting for a second or more and continuously decreasing in strength. The factors, which most strongly affect the structural response, are the inertia of the structure, as measured by the mass and strength of the structure, the overall structural design and the ductility of the materials and members comprising the structure. The strength of the structure is not only a material characteristic but includes the massiveness of the construction and redundancy of supports. If the strength is not isotropic, then the orientation of the structure with respect to the burst becomes important [Kinney & Graham, 1985; Glasstone & Dolan, 1987; Bailey & Murray, 1989].

Impact, by contrast, is a localized phenomenon. Upon collision with a target, the process of transforming some or all of the kinetic energy of the striker into deformation and failure of the colliding materials is governed by wave propagation. On contact, a compression wave propagates into both striker and target with an initial intensity of $\rho c v$, where c represents sound speed, a material characteristic, and v is the particle velocity. Because of geometric effects, this initial intensity rapidly decays to a value of $v^2/2$. The presence of free surfaces and material interfaces causes multiple reflections of the propagating waves causing the occurrence of sharp compressive and tensile stress gradients. Depending on the duration of the tensile pulses and their intensity, material failure in the form of physical separation of material may occur by a number of mechanism (see Curran, Seaman & Shockey [1977]; Backman & Goldsmith [1978]; Johnson [1972]; Zukas *et al.* [1982]; Zukas [1987, 1990]; Bushman *et al.* [1993]; Blazynski [1987] and Jones [1989]; Walters and Zukas [1989] for comprehensive discussions on impact phenomena).

The stress waves in the colliding solid propagate at a characteristic velocity that is a property of the material. The particle velocity (the speed at which material particles are actually displaced in the solid), however, governs the ensuing deformation. Because of this, coupled with the presence of multiple wave reverberations due to geometric effects, the most intense deformations occur within 3–6 characteristic striker dimensions (e.g., 3–6 diameters if the striker has regular geometry such as a cylinder, sphere or cone). Within this region, strain rates in excess of 10^5 are not uncommon, as are plastic strains exceeding 60% and pressures well in excess of the material strength. Loading and response times are in the sub-millisecond regime. Unlike structural response problems the high-frequency components of the structure are excited and lead to material failure by a wide variety of mechanisms.

Because of the localized nature of the impact processes [Wright & Frank, 1988], the geometries that need to be considered can be somewhat simplified. Unlike the structural dynamics problem, catastrophic failure by penetration or perforation can occur long before the distant boundaries have any effect. Thus, boundary conditions tend to be very simple, allowing either for perfect transmission or reflection of the incoming signal. Initial conditions most often specify the velocity of the striker, although these can be augmented by surface pressures or shears depending on the problem.

When it becomes necessary to consider the lethal effects of debris caused by blast loading, the blast and impact problems share several common features. Wave propagation effects must be accounted for explicitly since the interaction of stress waves with free surfaces, material interfaces and geometric discontinuities lead to material failure such as spallation or fragmentation. Inertia effects must be accounted for

150   *Introduction to Hydrocodes*

in considering both types of problems. This implies the use of explicit time integration techniques in numerical analyses as well as contact–impact (sliding interface) logic to properly account for momentum distribution between flying debris and surrounding structures. Both types of problems need constitutive models for deformation and failure at high strain rates. The material data or constants, which are embedded in these models must also be determined from wave propagation experiments, i.e., at strain rates appropriate to the problems being analyzed. Hence, the early stages of a problem in blast loading and most high-velocity impact problems can be successfully analyzed by wave propagation codes, often referred to as hydrocodes.

a)   *Constitutive descriptions for metallic materials.*

Modern hydrocodes decouple the material response for metallic materials into a volumetric and a deviatoric part. The volumetric or hydrodynamic response is obtained through an EOS. By far the most popular EOS in existing codes is the Mie–Gruneisen EOS, which makes use of Hugoniot data obtained from plate impact experiments and includes an energy term to reach off-Hugoniot states. The Mie–Gruneisen EOS assumes that the material will respond as a solid throughout its loading regime. For transitions from solid to vapor or liquid states, another equation must be used. The Tillotson EOS has been very popular for many years for problems in hypervelocity impact where such transitions occur with high frequency. Examples of equations of state in modern hydrocodes are shown in Figs. 4.5 and 4.6. Extensive compilations of EOS data exist, due to van Thiel [1977], Marsh [1980], Dobratz [1981], Kohn [1969], Vigil [1989] and others. The outstanding text by Drumheller [1998] should be read for more information on EOS formulations and the use of equations of state in hydrocodes. For situations where EOS data does not exist, there are at least a dozen methods by which it can be obtained. Thus, years of experimentation and systematic collection of shock wave data have made computation of the hydrodynamic response of materials almost a routine matter. Such is not the case when considering material strength, or deviatoric, effects.

An incremental elastic–plastic formulation is used to describe the shear response of metals in present finite-difference and finite-element codes. The plasticity descriptions are based on the assumed decomposition of the velocity strain tensor, $\dot{\varepsilon}$, into elastic and plastic parts together with incompressibility of the plastic part

$$\dot{\varepsilon} = \dot{\varepsilon}^e + \dot{\varepsilon}^p. \tag{4.10}$$

$$\dot{\varepsilon}^p_{11} + \dot{\varepsilon}^p_{22} + \dot{\varepsilon}^p_{33} = 0. \tag{4.11}$$

The von Mises yield criterion is typically used to describe the onset of yielding. Provision is made to account for strain hardening, compressibility and thermal effects but more often than not such data is not available for practical calculations. Earlier hydrocodes incorporated the Jaumann stress rate. Because of difficulties with this formulation at strains exceeding 40% [Walters & Zukas, 1989], this formulation is gradually being replaced with alternatives [Green & Naghdi, 1965; Dienes, 1979]. The typical constitutive model for metallic materials is depicted in Figs. 4.3 and 4.4.

Dynamic events often involve increases in temperature due to adiabatic heating. To accurately predict the response of a material, the effect of temperature on the flow stress must be included in a constitutive model. An attempt to account for the effects of strain, strain rate and temperature is made in the model proposed by Johnson & Cook [1983, 1985]. This is an empirical model involving five constants A, B, C, n and m. The functional form

$$\sigma = [A + B\varepsilon^n][1 + C \ln \dot{\varepsilon}^*][1 - (T^*)^m], \tag{4.12}$$

where

$$T^* = \frac{(T - T_{\text{room}})}{(T_{\text{melt}} - T_{\text{room}})}, \tag{4.13}$$

and

$$\dot{\varepsilon}^* = \frac{\dot{\varepsilon}}{\dot{\varepsilon}_0}, \quad \dot{\varepsilon}_0 = 1.0 \text{ s}^{-1}, \tag{4.14}$$

of the model constitutes the authors' best guess as to how material behaves under high-rate loading. It is not based on any theory. The constants must be determined from experiments. Both the experiments and the model decouple the effects of the parameters, which enter into it. One result is the implication of the model that strain rate sensitivity is independent of temperature, a feature that is not generally observed for most metals [Nicholas & Rajendran, 1990]. In fact, rate sensitivity is found to increase with increasing temperature while flow stress decreases.

Zerilli & Armstrong [1987] used dislocation dynamics concepts which also accounts for strain, strain rate and temperature effects in a coupled manner. For face-centered cubic (f.c.c.) metals their model takes the form

$$\sigma = C_0 + C_2\varepsilon^n[e^{(-C_3T+C_4T \ln \dot{\varepsilon})}], \tag{4.15}$$

while for body-centered cubic (b.c.c.) metals it is given as

$$\sigma = C_0 + C_1[e^{(-C_3 T + C_4 T \ln \dot{\varepsilon})}] + C_5 \varepsilon^n, \tag{4.16}$$

where σ represents stress, ε and $\dot{\varepsilon}$ represent strain and strain rate, respectively, T is the temperature and the Cs are material constants. The rationale for the difference in the two forms is in the stronger dependence of yield stress on temperature and strain rate is known to result for b.c.c. metals than f.c.c. metals. The constants for the model have been determined for only two materials. A comparison was made between the Johnson–Cook (JC) and Zerilli–Armstrong models for Taylor cylinder impact experiments [Johnson & Holmquist, 1988]. The Zerilli–Armstrong model gave slightly better correlation with experimental results although, being small strain models, neither was very accurate in describing large strain behavior [Nicholas & Rajendran, 1990].

A problem with the Johnson–Cook model as given above is that it does not treat the sudden strengthening that many ductile metals exhibit at strain rates greater than 10^4 s^{-1}. Rule & Jones [1998] propose a revised form of the Johnson–Cook to take this sudden increase in yield stress into account

$$\sigma = (C_1 + C_2 \varepsilon^N)\left[1 + C_3 \ln \varepsilon^* + C_4\left(\frac{1}{C_5 - \ln \varepsilon^*} - \frac{1}{C_5}\right)\right](1 - T^{*M}). \tag{4.17}$$

The strain rate sensitivity has been enhanced by the term

where C_5 is the natural logarithm of a critical strain rate level. This term tends to infinity as strain rate approaches the critical strain rate. Due to the $1/C_5$ correction term, the strain rate sensitivity enhancement term tends to zero for low strain rates. Thus, this revised, or RJC, model approaches the original JC model

$$\frac{1}{(C_5 - \ln \varepsilon^*)}, \tag{4.18}$$

for low strain rates and is identical to the JC model for a strain rate of unit magnitude where $\ln \varepsilon^* = 0$. Rule and Jones also propose a method to economically estimate all eight coefficients of the revised strength model using quasi-static tension data and Taylor impact test data.

The Bodner–Partom (B–P) model [1975] is based on dislocation dynamics concepts and treats strain rate and temperature effects in a coupled manner to more realistically account for observed behavior. It has been used in analysis of rate history effects [Bodner & Merzer, 1978], propagation of plastic waves in long rods [Bodner & Aboudi, 1983] and precursor decay analysis [Nicholas, Rajendran & Grove, 1987a,b].

The Bodner–Partom Model is shown in Figs. 4.39 and 4.40. Procedures for determining the constants are given by Rajendran [1994]. Figure 4.40 also illustrates a problem with these models—the paucity of data, or constants, to drive the models. Data is usually available for the more commonly used alloys of aluminum, steel, copper, tungsten and uranium. Go much further afield and a material characterization program must accompany any computational effort.

Models developed by Steinberg, Cochran & Guinan [1987] and Steinberg [1987, 1991], assume strain rate saturation and allow variation of shear modulus and yield strength with pressure and temperature:

$$G = G_0\left[1 + \left[\frac{G'_p}{G_0}\right]\frac{p}{\eta^{1/3}} + \left[\frac{G'_T}{G_0}\right](T - 300)\right], \qquad (4.19)$$

$$Y = Y_0(1 + \beta\varepsilon)^n\left[1 + \left[\frac{Y'_p}{Y_0}\right]\frac{p}{\eta^{1/3}} + \left[\frac{Y'_T}{Y_0}\right](T - 300)\right], \qquad (4.20)$$

Steinberg [1987] catalogs the available parameters for the model.

Other models which account for the variation of strength with strain, strain rate, temperature, pressure or other parameters are discussed in the literature [Nicholas & Rajendran, 1990; Rajendran, 1994; Follansbee and Kocks, 1988; Meyer, 1992]. These are generally 1D models and have been used to interpret various experiments involving high rate behavior. In principle, 1D models can be generalized to three dimensions through the use of effective stress and strain formulations. Except for isolated incidences to support specific laboratory experiments, these have not found their way into production computer codes and will not be mentioned further.

$$\dot{\varepsilon}^p_{ij} = D_0\, e^{\left(-\frac{(n+1)}{n}\left(\frac{Z^2}{3J_2}\right)\right)^n} \frac{S_{ij}}{\sqrt{J_2}}.$$

$$\dot{Z} = m(Z_1 - Z)\dot{W}_p.$$

D_0 is the limiting strain rate, n is strain rate sensitivity parameter
m is strain hardening parameter
Z_0 and Z_1 are initial and saturated values of the internal state variables Z
Z describes resistance to plastic flow, and loading history dependency.
n is a temperature dependent constant. W_p is plastic work.

Figure 4.39 Bodner–Partom model.

Material	Z_0 (GPa)	Z_1 (GPa)	n	m_0 GPa^{-1}	m_1 GPa^{-1}	α GPa^{-1}	A	B
C1008 Steel	5.5	7.0	0.4	15	0	0	0.245	46
HY100 Steel	2.4	3.6	1.2	10	0	0	NA	NA
1020 Steel	0.64	0.93	4.0	30	0	0	NA	NA
MAR-200 Steel	2.2	2.4	4.0	5	0	0	NA	NA
Armco Iron	2.65	4.2	0.58	56	0	0	NA	NA
OFHC Copper	0.8	6.6	0.4	11	150	1500	NA	NA
6061-T6 Aluminum	0.45	0.55	4.0	120	0	0	-2.86	2343
7039-T64 Aluminum	0.56	0.76	4.0	28	0	0	NA	NA
Pure Tantalum	1.3	3.1	0.74	20	0	0	NA	NA
W-2 Tungsten	8.75	10.0	0.58	150	0	0	0.166	134
Nickel 200	0.32	0.82	4.0	40	0	0	NA	NA
MAR-250 Steel	2.5	2.7	5.0	20	0	0	NA	NA
AF1410 Steel	2.4	2.75	5.0	15	0	0	NA	NA

NA -- The high temperature constants are "Not Available"

Figure 4.40 Bodner–Partom model constants.

Material failure is treated as an instantaneous process, based on maxima or minima of field variables (see Zukas [1987] for detailed descriptions of hydrocode failure criteria and post-failure models). It is well known at this time that material failure is a time-dependent phenomenon. However, the degree of material characterization required for advanced failure models taking into account nucleation, growth and coalescence of microvoids and cracks to form free-flying fragments is so expensive that it is not used in practice. Simple, one-parameter failure criteria are known to be inadequate descriptors of material failure at high strain rates. However, they frequently provide answers which are close enough for all practical purposes and the parameters for such models can frequently be obtained from existing high-rate data.

The above formulations are virtually common to all wave propagation codes. In addition, some codes also allow for a number of alternative material descriptions: elastic (isotropic or orthotropic); piecewise linear elastic–plastic models with strain hardening effects; linear viscoelastic; thermo-elastic; thermo-plastic with material properties temperature-dependent. In addition to the von Mises flow rule, general isotropic, kinematic and combined strain hardening rules allow yield surface translation as well as expansion or contraction.

It is generally agreed [Immele, 1989] that hydrocodes are a vital supplement to impact experiments. Computations generally return more insight and data than tests; allow generalizing the test data to assess the effects of variables (e.g., strength, detonation velocity) that had not been tested and permit parameter variations and tests of sensitivity. Organizations making extensive use of numerical simulations find that the number of tests can often be reduced and that fewer surprises are encountered during testing, all resulting in reduced costs and manpower demands [Herrmann, 1977; Immele, 1989]. The single biggest drawback to the accuracy of code predictions is the primitive status of the failure models [NMAB, 1980; Zukas, 1987]. If inertia effects dominate material or structural response, code predictions can be quite accurate—with 3–5% of experimentally observed deformations, penetration depths and residual velocities. When brittle or ductile failure modes dominate material response, code results are very likely to be wrong in both qualitative and quantitative predictions of impact events.

b)   *Constitutive descriptions for non-metallic materials.*

A large number of practical problems involve impact or impulsive loading of structures consisting primarily of non-metallic materials. Reactor containment structures must withstand external impacts of commercial aircraft as well as tornado-borne debris such as automobiles, telephone poles, rocks and other readily available natural and manmade materials. Such structures are typically made of reinforced concrete or layers of concrete and steel. Protective bunkers for personnel or storage facilities for energetic materials are often made of layers of metal, concrete and soil. Body armor consists of multiple layers of Kevlar or nylon. Impacts onto sand, water, ice, snow and various geological materials are not uncommon in everyday applications. Ceramics are being used in many industrial and military applications. Finally, there is a very large class of composite materials that is slowly beginning to take the place of metals, if not for ballistic protection, then for safety considerations. Perforation of a metallic fuel tank on an aircraft may cause failure through a hydrodynamic ram effect. However, if the aircraft survives that, sparks caused by penetration of a metallic tank by a metallic striker can cause explosion of the fuel. Replacement of the fuel tank and other aircraft components by composite materials with strengths similar to that of metals will not improve ballistic performance but will inhibit sparks [Held, 1993].

What, then, is the status of computations with such materials? Thoma & Vinckier [1993] summarize the situation very nicely: "...whereas the calculations of high velocity impact on isotropic materials can be done on a routine basis, the simulation of the impact and penetration process into nonisotropic materials like reinforced concrete or fiber-reinforced materials is still a research task." This state of affairs results from a lack of:

(a) Definitive computational models.

(b) High-strain-rate data for both volumetric and deviatoric response for anisotropic materials.

(c) An experimental database resulting from controlled, instrumented impact experiments to serve as a basis for code validation.

The first of these issues is probably the easiest to resolve. A short description of constitutive models for non-metals is given next. Ceramics are specifically excluded here since this is an area of active and sometimes contradictory research due to a paucity of data.

Soils and crushable (porous) materials are frequently represented by a p-alpha model first developed by Herrmann [1969]. A variant of this formulation is incorporated in the NIKE code (see Zukas [1990] for a review of modern hydrocodes). ADINA includes a concrete model with multi-axial material failure envelopes on tension cracking and compression crushing and strain softening; a Drucker–Prager type model for soils and rocks with tension cutoff and compression cap; and a curve (empirical) description soil and rock model with tension cutoff or cracking. A version for plain concrete is included in the EPIC and HULL codes. Burton & Schatz [1975] used such a model in the TENSOR to study rock crushing.

Data for a p-alpha type model must be determined from individual materials from a number of independent experiments. The generic form of such a model is depicted in Fig. 4.41. The material originally loads along the virgin path A. If the maximum compression is less than ζ_m, unloading also occurs along path A. A totally crushed material would unload along path B. For partially crushed materials the unloading path would be C unless there is elastic recovery, in which case the unloading path becomes D. In order to create such a pressure–compression map, the points ζ_1, ζ_2, ζ_3 and the residual porosity α must be determined through laboratory tests, which can involve the iterative use of computations with an assumed constitutive model and indirect experiments.

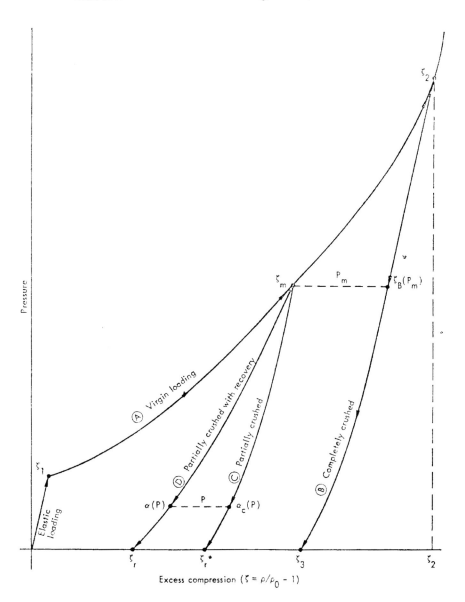

Figure 4.41 Constitutive model for porous materials.

Geological materials are commonly handled through a CAP model [Gupta & Seaman, 1979; Wright & Baron, 1979], shown in Fig. 4.42. This has also been adapted for porous materials, concrete and ceramics by judicious adjustment of the various parameters. A non-associative flow rule is commonly used to describe the shear envelope. The cap moves depending on the degree of volumetric response.

158 *Introduction to Hydrocodes*

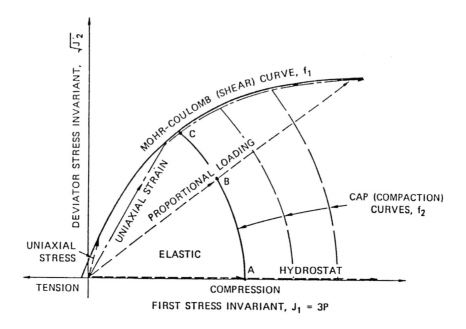

Figure 4.42 CAP model.

Many porous materials yield under shear loading, compact irreversibly under compressive loading, and separate in tension. In concrete and rocks, the shear strength increases with increasing pressure. Also, increasing shear stress reduces the resistance to compaction. This shear-enhanced compaction is common to porous materials, the stiffness of the material also increases with pressure; hence the elastic moduli are functions of the stress state. Thus, CAP models generally include the following:

(a) Elastic stress–strain relations, including variation of the moduli from initial loading to consolidation.

(b) Mohr–Coulomb (or shear) curve, including plastic stress–strain relations on the curve.

(c) CAP curve, including plastic stress–strain relations on the CAP curve and description of the strain hardening of the CAP curve.

(d) Tensile curve, which is the continuation of the Mohr–Coulomb curve into the tension region, stress–strain relation and stress reduction on the curve, and a fracture and separation process.

A comprehensive description of the derivation of a CAP-type model for reinforced concrete as well as experiments to determine the necessary model parameters are given in the report by Gupta & Seaman [1979].

The most complete model for homogeneous orthotropic materials has been incorporated in EPIC. Elastic behavior is determined from the anisotropic elastic stress–strain relationship

$$\sigma_{ij} = C_{ijkl}\varepsilon_{kl}, \tag{4.21}$$

where C_{ijkl} is specialized to an orthotropic material. The onset of plasticity is determined by the Hill criterion

$$\sigma_{\text{eff}} = \left[\frac{H_1}{2}(\sigma_x - \sigma_y)^2 + \frac{H_2}{2}(\sigma_x - \sigma_z)^2 + \frac{H_3}{2}(\sigma_y - \sigma_z)^2 + 3F_{44}T_{xy}^2 \right. \\ \left. + 3F_{55}T_{zx}^2 + 3F_{66}T_{yz}^2 \right]^{1/2}, \tag{4.22}$$

which reduces to the von Mises law when anisotropy is vanishingly small. The plasticity model also accounts for anisotropic work-hardening. The drawback is that the H and F parameters must be determined by the user. Aside from the difficulty of performing experiments to determine six free parameters, there is also that comment by Cauchy as to what you can do with them. Pressure is determined from either a polynomial EOS fit to data for the specific composite or estimated from an isotropic Mie–Gruneisen EOS. Complete derivation of the equations and the numerical algorithm are described by Johnson, Vavrick & Colby [1980].

An alternate approach to determination of the high-pressure response using mixture theory has been proposed by Anderson, O'Donoghue & Skerhut [1990]. Anisotropic constitutive formulations offering minor corrections to the EPIC approach have been proposed by O'Donoghue et al. [1992] and Anderson et al. [1994].

The anisotropic models in other hydrocodes tend to model only elastic behavior, typically viewing the composite as homogeneous but transversely isotropic or orthotropic. Isotropic behavior is assumed upon the onset of plasticity. For examples see Hallquist [1986], Swegle & Hicks [1979], Key, Beisinger & Krieg [1978], Sedgwick et al. [1976] and Hallquist & Whirley [1989] to cite but a few.

The above develop models for composites which retain their full load-carrying capability until failure. By and large, however, there is no agreement as to what constitutes failure nor how the material should be treated after failure has occurred. Keep in mind that the above models all assume *homogeneous anisotropy*. As much or more work remains to be done here regarding failure of composites under high-rate loading as for metallic materials. See Abrate [1998] and Sierakowski & Chaturvedi [1997] for the current status of dynamic loading and characterization of composites.

In addition, a large number of *ad hoc* models exist for solids, liquids and explosives. These usually lack theoretical foundation and are little more than attempts to fit existing and incomplete databases. As many successful calculations with such models can be found as unsuccessful ones. The current interest is in computational models for ceramics and for reinforced concrete. At present, no definitive computational models have emerged. However, this is an area of active research.

4.7   Problem areas.

Assuming that the above models prove adequate in the short run for non-metallic materials—a very dangerous assumption—what about the remaining problems? Where is the data to drive these models? What experiments are available to validate code predictions?

Most of the data available so far comes from traditional high-rate test techniques such as the split-Hopkinson pressure bar, the expanding ring test and plate impact experiments. See Nicholas & Rajendran [1990] for a description of high-strain-rate test methodology. All such data, however, assesses uniaxial behavior. Biaxial and triaxial evaluation techniques are still in their infancy, incapable of generating on a production basis the data required for routine calculations. The problem can be seen from the following example.

A manufacturer develops a new fiber. A load–elongation curve is determined along the fiber axis but no transverse measurements are made. The fiber is embedded in an isotropic matrix. By alternating lay-ups, a composite structure of some sort is built. At best, a few static measurements are made on the constituents.

Assuming the resulting structure, be it beam, cylinder, plate or whatever, is orthotropic, nine constants are required to characterize its elastic behavior. Another six are required for the anisotropic yield criterion. At least one parameter is required to characterize work-hardening. Assuming that a one-term EOS will suffice, another constant is required here. This totals 17 required material descriptors, all determined at strain rates appropriate to the problem at hand. At best, two are available from static tests. Allowing for Divine Intervention, say all nine parameters for the elastic behavior are available, but from static tests results. Now only eight more are required, plus an answer to the question of whether static data for composites is a valid descriptor of high rate behavior (our experience with metals indicates that in general the answer will be no).

In order to complete the calculation, the remaining parameters must be determined in some fashion. To assume that this will be done in a logical and consistent manner, following the scientific method, is to place more faith in computational specialists than has been heretofore been warranted. Thus, when evaluating results, in mind the following comment as quoted by Glenn & Jannach [1977]: "...we were recently reminded of a comment attributed to the French mathematician, Cauchy, to the effect that, given four parameters, he could draw a credible version of an elephant and given five, he could make its tail wiggle."

Assuming a credible calculation can be completed, how are the results to be displayed? For isotropic materials the important parameters governing short-term response of materials and structures to intense loading are well known. The picture is not as clear for non-metallic, anisotropic materials. The codes can and do generate volumes of data, but which of the data is significant for particular applications is a matter of judgment. Finally, how will the validity of the calculation be assessed? Comparison of computed results with global parameters such as gross deformations, energies, momenta and quantities related to these are necessary but not sufficient conditions. For metallic materials they are not adequate discriminants of the validity of a constitutive formulation. For non-metallic materials the case is likely to be worse. For metallic materials, experimental procedures exist for *in situ* measurement of arrival times, pressures, wave profiles, strains and other local quantities. For most non-metallic materials, such experiments remain to be defined.

The principal experiments which are used to generate high-strain-rate data are discussed in Chapter 7. Representative results are presented as well.

4.8   Summary.

Hydrocodes, or wave propagation codes, are a valuable adjunct to the study of the behavior of metals subjected to high-velocity impact or intense impulsive loading. The combined use of computations, experiments and high-strain-rate material characterization has, in many cases, supplemented the data achievable by experiments alone at considerable savings in both cost and engineering man-hours. A large database exists of high-pressure EOS data. Considerable data on high rate deviatoric behavior exists as well although, unlike EOS data, it is not collected in a few compilations but scattered throughout a diverse literature. Experimental techniques exist for determining either EOS or strength data for materials not yet characterized under high-rate loading conditions.

By contrast, computations with non-metallic materials such as composites, concrete, rock, soil and a variety of geological materials are, in effect, research tasks. This is due to several reasons: lack of definitive computational models for high strain rate–temperature–pressure response; lack of a database for EOS and high rate strength data for such materials; lack of test methodologies for anisotropic materials subjected to high-rate loading.

A large number of *ad hoc* models exist for explosives, geological materials, concrete and other non-metallics. Many of these lack a firm theoretical foundation. This is an area where considerable research is required, both to devise appropriate test techniques to measure material response under high strain rates, elevated temperatures and high pressures as well as to develop appropriate constitutive models.

REFERENCES

NMAB. (1980). Materials response to ultra-high loading rates, Report NMAB-356, National Materials Advisory Board, Washington, DC.

ABAQUS. (2001). Troubleshooting ABAQUS/explicit analysis: difficulties with contact, ABAQUS website, *http://www.hks.com/support/troubleshooting/ts_exp_contact_issues.html*.

S. Abrate. (1998). *Impact on Composite Structures*. Cambridge: Cambridge U. Press.

M.H. Aliabadi and C.A. Brebbia (Eds.) (1993). *Contact Mechanics: Computational Techniques*. Southampton, U.K.: Computational Mechanics Publications.

M.H. Aliabadi and C. Alessandrini (Eds.) (1995). *Contact Mechanics II: Computational Techniques*. Southampton, U.K.: Computational Mechanics Publications.

C.E. Anderson Jr., P.E. O'Donoghue and D. Skerhut. (1990). A mixture theory approach for the shock response of composite materials. *J. Comp. Mater.*, 24, 1159-1178.

C.E. Anderson Jr., P.A. Cox, G.R. Johnson and P.J. Maudlin. (1994). A constitutive formulation for anisotropic materials suitable for wave propagation computer programs—II. *Comput. Mech.*, 15, 201-223.

M.E. Backman and W. Goldsmith. (1978). The mechanics of penetration of projectiles into targets. *Int. J. Engng Sci.*, 16, 1-99.

A. Bailey and S.G. Murray. (1989). *Explosives, Propellants and Pyrotechnics*. London: Brassey's.

D. Baraff. (1989). Analytical methods for dynamic simulation of non-penetrating rigid bodies. *Comput. Graphics*, 23(3), 223-231.

D. Baraff. (1994). Fast contact force computation for non-penetrating rigid bodies. *Comput. Graphics (Proc. SIGGRAPH)*, Orlando, July 24–29, pp. 23-34.

D. Baraff. (1996). Linear-time dynamics using Lagrange multipliers. *Comput. Graphics (Proc. SIGGRAPH)*, New Orleans, August 4–9, pp. 137-146.

K.-J. Bathe. (1982). *Finite Element Procedures in Engineering Analysis*. Englewood Cliffs, NJ: Prentice-Hall.

T. Belytschko and T.J.R. Hughes. (1983). *Computational Methods for Transient Analysis*, Amsterdam: North-Holland.

T. Belytschko and J.I. Lin. (1985). A new interaction algorithm with erosion for EPIC-3, U.S. Army Ballistic Research Lab., Report BRL-CR-540, February.

T. Belytschko and W.K. Liu. (1983). On mesh stabilization techniques for under-integrated elements. In J. Chandra and J.E. Flaherty, eds, *Computational Aspects of Penetration Mechanics*. Heidelberg: Springer-Verlag.

T. Belytschko and M.O. Neal. (1989). The vectorized pinball contact impact routine, in A.H. Hadjian, ed, *Trans. 10th Int. Conf. on Structural Mechanics in Reactor Technology*, Anaheim, CA.

T. Belytschko, J. Ong and W.K. Liu. (1984). Hourglass control in linear and nonlinear problems. *Comput. Methods Appl. Mech. Engng*, 43(3), 251-276.

D.J. Benson. (1992). Computational methods in Lagrangian and Eulerian hydrocodes. *Comput. Methods Appl. Mech. Engng*, 99, 235-394.

T. Blazynski (ed). (1987). *Materials at High Strain Rates*. London: Elsevier Applied Science.

S.R. Bodner. and A. Merzer. (1978). Viscoplastic constitutive equations for copper with strain rate history and temperature effects. *Trans ASME, J. Engng. Mat. Tech.*, 100, 388-394.

S.R. Bodner and J. Aboudi. (1983). Stress wave propagation in rods of elastic-viscoplastic materials. *Int. J. Solids Struct.*, 19, 305-314.

S.R. Bodner and Y. Partom. (1975). Constitutive equations for elastic-viscoplastic strain-hardening materials. *Trans. ASME, J. Appl. Mech.*, 42, 385-389.

C.A. Brebbia and V. Sanchez-Galvez (Eds.) (1994). *Shock and Impact on Structures*, Southampton, U.K.: Computational Mechanics Publications.

R.F. Breidenbach, S.R. Jones, V.M. Smith and B.L. Yaney. (1997). Finite element contact algorithms in k-DYNA 3D suitable for the high speed impact and penetration problems. In Y.S. Shin, J.A. Zukas, H.S. Levine and D.M. Jerome, eds, *Structures Under Extreme Loading Conditions 1997*, PVP-vol. 351. New York: ASME.

P.S. Bulson (Ed.) (1992). *Structures Under Shock and Impact II*. Southampton, U.K.: Computational Mechanics Publications.

P.S. Bulson (Ed.) (1994). *Structures Under Shock and Impact III*. Southampton, U.K.: Computational Mechanics Publications.

D.E. Burton and J.F. Schatz. (1975). Rock modeling in TENSOR74, a two-dimensional Lagrangian shock propagation code, Report UCID-16719, Lawrence Livermore Laboratory.

A.V. Bushman, G.I. Kanel', A.L. Ni and V.E. Fortov. (1993). *Intense Dynamic Loading of Condensed Matter*. Washington, DC: Taylor & Francis.

E.P. Chen and V.K. Luk (Eds.) (1993). *Advances in Numerical Simulation Techniques for Penetration and Perforation of Solids*. AMD-Vol. 171, New York: ASME.

R.D. Cook. (1995). *Finite Element Modelling for Stress Analysis*, NY: Wiley.

D.R. Curran, L. Seaman and D.A. Shockey. (1977). Dynamic failure in solids. *Phys. Today*, 30, 46-55.

R. Diekmann, J. Hungershofer, M. Lux, L. Taenzer and J.-M. Wierum. (2000). Efficient contact search for finite element analysis, in *Proc. European Congress on Computational Methods in Applied Sciences and Engineering, ECCOMAS 2000*, Barcelona, Spain. 11–14 Sept. 2000.

J.K. Dienes. (1979). On the analysis of rotation and stress rate in deforming bodies. *Acta Mech.*, 32, 217ff.

B.M. Dobratz. (1981). LLNL explosives handbook, Report UCRL-52997, Lawrence Livermore National Laboratory.

D.S. Drumheller. (1998). *Introduction to Wave Propagation in Nonlinear Fluids and Solids*. Cambridge: Cambridge University Press.

D.P. Flanagan and T. Belytschko. (1981). A uniform stain hexahedron and quadrilateral with orthogonal houglass control. *Int. J. Num. Methods Engng*, 17, 679-706.

P.S. Follansbee and U.F. Kocks. (1988). A constitutive description of the deformation of copper based on the use of mechanical threshold stress as an internal state variable. *Acta Metall.*, 36, 81ff.

I. Fried. (1979). *Numerical Solution of Differential Equations*. New York: Academic Press.

S. Glasstone and P.J. Dolan. (1977). *The Effects of Nuclear Weapons*, 3rd edn. Washington, DC: U.S. Govt. Printing Office.

L.A. Glenn and W. Jannach. (1977). Failure of granite cylinders under impact loading. *Int. J. Fract.*, 13, 301-317.

G.L. Goudreau and J.O. Hallquist. (1982). Recent developments in large-scale finite element Lagrangian hydrocode technology. *Comput. Methods Appl. Mech. Engng*, 33(1–3), 725-757.

A.E. Green and P.M. Naghdi. (1965). General theory of an elastic–plastic continuum. *Arch. Rat. Mech. Anal.*, 2, 197ff.

Y.M. Gupta and L. Seaman. (1979). Local response of reinforced concrete to missile impact, Report EPRI-NP-1217, Electric Power Research Institute.

J.O. Halliquist. (1976). A procedure for the solution of finite deformation contact–impact problems by the finite element method, Lawrence Livermore National Laboratory, Report UCRL-52066.

J.O. Hallquist. (1978). A numerical treatment of sliding interfaces and impact. In K.C. Park and D.K. Gartling, eds, *Computational Techniques for Interface Problems*, AMD-vol. 30. New York: ASME.

J.O. Hallquist. (1986). NIKE2D—A vectorized, implicit, finite deformation, finite element code for analyzing the static and dynamic response of 2D solids with interactive rezoning and graphics, Report UCID-19677, Rev. 1, Lawrence Livermore National Laboratory.

J.O. Hallquist. (1988). User's manual for DYNA2D—an explicit two-dimensional hydrodynamic finite element code with interactive rezoning and graphical display, Lawrence Livermore National Laboratory, UCID-18756, Rev. 3, February.

J.O. Hallquist and R.G. Whirley. (1989). DYNA3D user's manual, Report UCID-19592, Rev. 5, Lawrence Livermore National Laboratory.

J.O. Hallquist, G.L. Goudreau and D.J. Benson. (1985). Sliding interfaces with contact–impact in large-scale Lagrangian computations. *Comput. Methods Appl. Mech. Engng*, 51(107).

M. Held. (1993). Fragment threat against aircraft structures, Proceedings of the Symposium on Transport Aircraft Survivability. Washington, DC: American Defense Preparedness Association.

W. Herrmann. (1969). Constitutive equations of the dynamic compaction of ductile porous materials. *J. Appl. Phys.*, 40(6).

W. Herrmann. (1977). Current problems in the finite difference solution of stress waves. In T.C.T. Ting, R.J. Clifton and T. Belytschko, eds, *Nonlinear Waves in Solids*, Army Research Office.

J.D. Immele (chmn.). (1989). Report of the review committee on code development and material modeling, Report LA-UR-89-3416, Los Alamos National Laboratory.

O.-P. Jaquotte and J.T. Oden. (1984). Analysis of hourglass instabilities and control in underintegrated finite elements. In W.K. Liu, T. Belytschko and K.C. Park, eds, *Proc. Intl Conf. on Innovative Methods for Nonlinear Problems*. Swansea, UK: Pineridge Press.

W. Johnson. (1972). *Impact Strength of Materials*. London: Edward Arnold.

G.R. Johnson. (1978). EPIC-2, a computer program for elastic–plastic impact computations plus spin, U.S. Army Ballistic Research Laboratory, Report ARBRL-CR-00373.

G.R. Johnson and W.H. Cook. (1983). A constitutive model and data for metals subjected to large strains, high strain rates and high temperatures. *Proc. 7th Intl Symposium on Ballistics*, The Hague, The Netherlands.

G.R. Johnson and W.H. Cook. (1985). Fracture characteristics of three metals subjected to various strains, strain rates, temperatures and pressures. *Engng Fract. Mech.*, 21, 31-48.

G.R. Johnson and T.J. Holmquist. (1988). Evaluation of cylinder-impact data for constitutive model constants. *J. Appl. Phys.*, 64, 3901-3910.

G.R. Johnson, D.J. Vavrick and D.D. Colby. (1980). Further development of EPIC-3 for anisotropy, sliding surfaces, plotting and material models, Report ARBRL-CR-00429, U.S. Army Ballistic Research Laboratory.

N. Jones. (1989). *Structural Impact*. Cambridge: Cambridge University Press.

S.W. Key, Z.E. Bensinger and R.D. Krieg. (1978). HONDO II: a finite element computer program for the large deformation dynamic response of axisymmetric solids, Report SAND78-0422, Sandia National Laboratories.

G.F. Kinney and K.J. Graham. (1985). *Explosive Shocks in Air*, 2nd edn. New York: Springer-Verlag.

B.J. Kohn. (1969). Compilation of Hugoniot equations of state, Report AFWL-TR-69-38, Air Force Weapons Laboratory.

A. Konter. (2000). *How to Undertake a Contact and Friction Analysis*. Glasgow: NAFEMS.

D. Kosloff and G.A. Frazier. (1978). Treatment of hourglass patters in low order finite element codes. *Int. J. Num. Anal. Methods Geomech.*, 2.

J.P. Lambert. (1978). A residual velocity predictive model for long rod penetrators, U.S. Army Ballistic Research Laboratories, Report ARBRL-MR-02828.

S.P. Marsh (ed). (1980). *LASL Shock Hugoniot Data*. Berkeley: Univ. of California Press.

L.W. Meyer. (1992). Constitutive equations at high strain rates. In M.A. Meyers, L.E. Murr and K.P. Staudhammer, eds, *Shock-Wave and High Strain Rate Phenomena In Materials*. New York: Marcel Dekker.

J.E. Mottershead. (1993). *Finite Element Analysis of Contact and Friction—A Survey*, Glasgow: NAFEMS, Ref: R0025.

M.O. Neal. (1989). Contact–impact by the Pinball Algorithm with penalty, projection and augmented Lagrangian methods, PhD Dissertation, Northwestern University.

T. Nicholas and A.M. Rajendran. (1990). Material characterization at high strain rates. In J.A. Zukas, ed, *High Velocity Impact Dynamics*. New York: Wiley-Interscience Chapter 3.

T. Nicholas, A.M. Rajendran and D.J. Grove. (1987a). Analytical modeling of precursor decay in strain-rate dependent materials. *Int. J. Solids Struct.*, 23, 1601-1614.

T. Nicholas, A.M. Rajendran and D.J. Grove. (1987b). An offset yield criterion from precursor decay analysis. *Acta Mech.*, 69, 205-218.

W. Noh. (1976). Lawrence Livermore Lab., Report UCRL-52112.

P.E. O'Donoghue, C.E. Anderson Jr., G.J. Friesenhahn and C.H. Parr. (1992). A constitutive formulation for anisotropic materials suitable for wave propagation computer programs. *J. Comp. Mater.*, 26, 1861-1884.

P. Papadopoulos and R.L. Taylor. (1993). A simple algorithm for the three-dimensional finite element analysis of contact problems. *Comp. Struct.*, 46, 1107-1118.

J.-P. Ponthot and D. Graillet. (1998). An efficient implicit scheme for the treatment of frictional contact between deformable bodies. In S. Idelsohn, E. Onate and

E. Dvorkin, eds, *Computational Mechanics: New Trends and Applications*. Barcelona: CIMNE, pp. 1-19.

A. Poth, et al. (1981). Experimental and numerical investigation of the ricochetting of projectiles off metallic surfaces, *Proc. 6th Intl Symp. on Ballistics*, Orlando, FL.

A. Poth, et al. (1983). Failure behavior of an aluminum plate under impact loading, *Proc. Intl. Conf. on Applications of Fracture Mechanics to Materials and Structures*, Freiburg, GE.

A. Poth, et al. (1985). Impact damage effects and computational methods, *AGARD, 60th Meeting of the Structures and Materials Panel*, San Antonio, TX.

A.M. Rajendran. (1994). Material constitutive models in short course notes. *Material Behavior at High Strain Rates*. Baltimore, MD: Computational Mechanics Associates.

Y.D.S. Rajapakse and J.R. Vison (Eds.) (1995). *High Strain Rate Effects on Polymer, Metal and Ceramic Matrix Composites and Other Advanced Materials*, AD-Vol. 48, New York: ASME.

W.K. Rule and S.E. Jones. (1998). A revised form for the Johnson–Cook strength model. *Int. J. Engng Sci.*, 21(8), 609-624.

J.C. Schulz. (1985). Finite element hourglassing control. *Int. J. Num. Methods Engng*, 21.

J.C. Schulz and O.E.R. Heimdahl. (1986). Hourglassing control through boundary smoothing. *Commun. Appl. Num. Methods*, 2.

R.T. Sedgwick, J.L. Waddell, L.J. Hageman, G.A. Gurtman and M. Baker. (1976). Influence of ABM material properties on erosion resulting from particle impact, Report SSS-R-76-2886, Systems, Science & Software, Inc.

S. Segletes and J.A. Zukas. (1996). The contact/erosion algorithm in the ZeuS hydrocode. In M.H. Aliabadi and C.A. Brebbia, eds, *Contact Mechanics: Computational Techniques*. Southampton: Computational Mechanics Publications.

Y.S. Shin (Ed.) (1995). *Structures Under Extreme Loading Conditions*, PVP-Vol. 299, New York: ASME.

Y.S. Shin and J.A. Zukas (Eds.). *Structures Under Extreme Loading Conditions–1996*, New York: ASME.

R.L. Sierakowski and S.K. Chaturvedi. (1997). *Dynamic Loading and Characterization of Fiber-Reinforced Composites*. New York: Wiley.

D.J. Steinberg. (1987). Constitutive model used in computer simulation of time-resolved shock wave data. *Int. J. Impact Engng*, 5(1–4), 603-612.

D.J. Steinberg. (1991). Equation of state and strength properties of selected materials, Report UCR-MA-106439, Lawrence Livermore National Laboratory.

D.J. Steinberg, S.G. Cochran and M.W. Guinan. (1987). A constitutive model for metals applicable at high strain rates. *J. Appl. Phys.*, 51, 1498-1504.

J.W. Swegle and D.L. Hicks. (1979). An anisotropic constitutive equation for use in finite difference wave propagation calculations, Report SAND79-0382, Sandia National Laboratories.

R.L. Taylor and P. Papadopoulos. (1993). On a finite element methods for dynamics contact/impact problems. *Int. J. Num. Methods Engng*, 36, 2123-2214.

K. Thoma and D. Vinckier. (1993). Numerical simulation of a high velocity impact on fiber-reinforced materials, *Impact IV: SMiRT Post-Conference Seminar*, Berlin.

M. van Thiel. (1977). Equation of state data, Report UCRL-50108, vols. 1–3, Lawrence Livermore National Laboratory.

J. von Neumann and R.D. Richtmyer. (1950). A method for the numerical calculations of hydrodynamical shocks, *J. Appl. Phys*, 23, 232-237.

B. Verhegghe and G.H. Powell. (1986). Control of zero-energy modes in a 9-node plane element. *Int. J. Num. Methods Engng*, 23.

M.G. Vigil. (1989). Projectile impact hugoniot parameters for selected materials, Report SAND89-1571, Sandia National Laboratories.

W.P. Walters and J.A. Zukas. (1989). *Fundamentals of Shaped Charges*. New York: Wiley-Interscience republished 1997 by CMC Press, Baltimore, MD.

M.L. Wilkins. (1980). Use of artificial viscosity in multidimensional fluid dynamics calculations. *J. Comput. Phys.*, 36, 281-303.

M.L. Wilkins. (1999). *Computer Simulation of Dynamic Phenomena*. New York: Springer-Verlag.

J.P. Wright and M.L. Baron. (1979). Dynamic deformation of materials and structures under explosive loading. In K. Kawata and J. Shioiri, eds, *High Velocity Deformation of Solids*. Berlin: Springer-Verlag.

T.W. Wright and K. Frank. (1988). Approaches to penetration problems. In W. Amman, W.K. Liu, J.A. Studer and T. Zimmerman, eds, *Impact: Effects of Fast, Transient Loading*. Rotterdam: A.A. Balkema.

F.J. Zerilli and R.W. Armstrong. (1987). Dislocation-mechanics-based constitutive relations for material dynamics calculations. *J. Appl. Phys.*, 61, 1816-1825.

J.A. Zukas. (1987). Fracture with stress waves. In T. Blazynski, ed, *Materials at High Strain Rates*. London: Elsevier Applied Science.

J.A. Zukas (ed). (1990). *High Velocity Impact Dynamics*. New York: Wiley-Interscience.

J.A. Zukas and B. Gaskill. (1996). Ricochet of deforming projectiles from deforming plates. *Int. J. Impact Engng*, 18(6), 601-610.

J.A. Zukas, T. Nicholas, H.F. Swift, L.B. Greszczuk and D.R. Curran. (1982). *Impact Dynamics*, New York: Wiley-InterscienceRepublished 1992 by Krieger Publishing Co., Malabar, FL.

Chapter 5
How Does a Hydrocode Really Work?

5.1 Introduction.

In the previous chapter we have covered the major ingredients of a wave propagation code. The preceding chapters were provided so you could understand the rationale behind the decisions that were made in adopting various numerical approaches and material descriptors. This material, and some of the suggested references, is sufficient to give you an overview of the theory of hydrocodes. Once you are familiar with the internal workings of a code, this is sufficient. On first approaching the subject, though, you may wonder how the puzzles are all put together. This is what we will cover next. In effect, we are going to go over some of the same material, but now from the point of view of the hydrocode. We take specifics from the *ZeuS* code, a Lagrangian, 2D explicit code for the study of fast, transient loading situations. Many of the approaches in *ZeuS* are similar to those found in leading Lagrange codes such as DYNA, EPIC, AUTODYN and PRONTO.

5.2 Pre-processing.

Figure 5.1(a) and (b) is taken from the *ZeuS* User's Manual. They show two-thirds of the process involved in running a hydrocode. The missing part deals with visualizing the results of a computation. This type of procedure is generic—you will find it in all codes, Euler or Lagrange. Some parts in Fig. 5.1(a) will change depending on whether the code uses finite differences or finite elements for discretization. Figure 5.1(b) is virtually identical for Lagrange codes. Look at EPIC and you will find almost the same diagram. Look at the manuals for DYNA, PRONTO, DYSMAS and you will find that the procedure is similar. For Euler codes, the process is a bit more complicated. For one thing, the box labeled "contact algorithm" will be replaced with one dealing with maintaining discrete interfaces. There will be an additional box dealing with material transport from cell to cell. However, the pressure, stress and force calculations will easily be recognizable as counterparts of the Lagrange set.

Let us start with Fig. 5.1(a). This is the part in the code where you specify the type of calculation you will do (1D, 2D, 2D plane strain or 3D), the geometry, the materials

170 *Introduction to Hydrocodes*

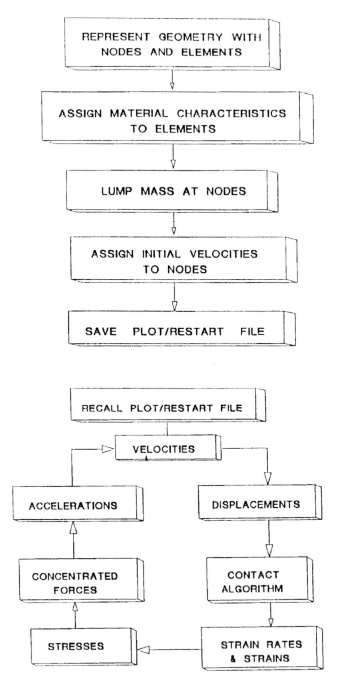

Figure 5.1 (a) The *ZeuS* code pre-processor. (b) Sequence of operations for *ZeuS* computations.

How Does a Hydrocode Really Work? 171

involved and their properties as well as the initial and boundary conditions. For example, if you do a 2D finite-element calculation, you would specify the boundaries of each of the bodies involved in a calculation, the type of elements to be used and the number of elements in each of the coordinate directions. In particular, say you are computing the impact of a right circular cylindrical projectile into a circular plate. You would specify the radius of the projectile and its length, the radial extent of the target and its thickness. Let us say for the sake of argument that the projectile has a radius of 0.5 cm and a length of 10 cm. Similarly, let the target extend radially to 10 cm and have a thickness of 1 cm. Further, let us quite arbitrarily specify the projectile–target interface to lie at the $z = 0$ coordinate line (Fig. 5.2). For an axisymmetric calculation, then, we specify projectile geometry to extend from $z = 0$ to $z = 10$ cm in length and from $r = 0$ to $r = 0.5$ cm in radius. The target, in turn, would go from $z = 0$ to $z = -1$ cm and from $r = 0$ to $r = 10$ cm.

Next we need to specify the number of cells or elements in the coordinate directions, depending on whether we are performing a finite-difference or finite-element spatial discretization. These are big words meaning that we will only compute results at

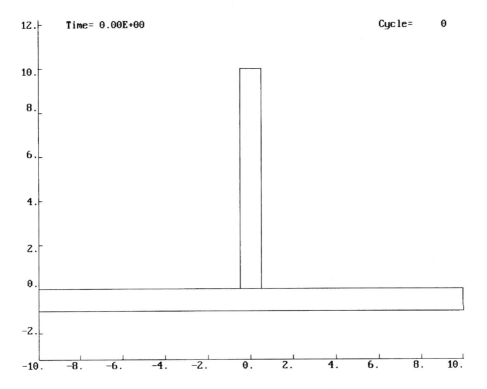

Figure 5.2 Outline of projectile impact calculation geometry.

a discrete number of points—the places where the cells and elements inter-connect. If we need information anywhere else in the grid, we will use our computed data at this finite number of points and interpolate.

For our example problem, let us determine that we want a uniform grid with each element being 0.1×0.1 cm^2. For the projectile, we therefore specify that we want six nodal points radially (at $r = 0.0$, 0.1, 0.2, 0.3, 0.4 and 0.50) and 101 points axially (starting at $z = 0$ and ending at $z = 10$ cm). Accordingly, the number of element layers will be 5 in the radial direction and 100 in the axial direction. Remember that quantities involving motion (acceleration, velocity and displacement) are computed at the nodes. Quantities related to deformation are computed at the element centroids (stresses, strains, strain rates, pressures, temperatures). Integral quantities such as energy, momentum and work are usually given for the body as a whole. If we use quadrilateral elements, our grid for the projectile consists of 5 radially × 100 axially = 500 *elements*. If we use triangular elements with four triangles per quadrilateral, the optimum arrangement, then we have $5 \times 100 \times 4 = 2000$ projectile elements. We do exactly the same thing for the target (Fig. 5.3).

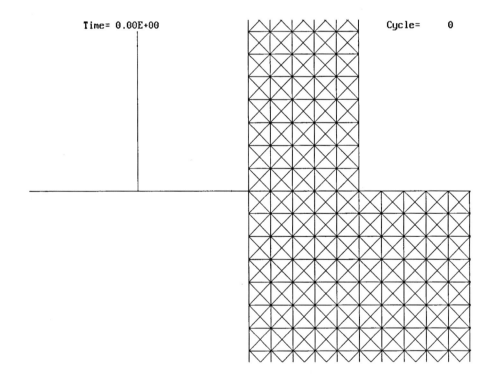

Figure 5.3 Detail of mesh for projectile impact calculation.

This discretization process is done for each body, or object, in the calculation. The actual work of specifying nodal and element coordinates is done by the geometry generator built into the code. It usually suffices to specify the outline of each object. The more flexible the geometry generator, the more useful the code.

At this point the geometry of the calculation has been defined. Next, it is necessary to specify:

(1) The materials of which the objects of a calculation are comprised and the properties (the constitutive model) for the material.

(2) The boundary conditions. (Fixed or free edges, added masses for wave propagation calculations. More elaborate descriptions are needed for structural dynamics codes.)

(3) The initial conditions (velocities, displacements, body forces, initial stresses or temperatures, surface loads).

(4) Contact conditions between the objects of the calculation.

(5) The initial time step, the maximum time step and the minimum time step which, if exceed, will cause the calculation to be stopped. This is for reasons of economy. A minuscule time step is usually an indication of a problem that may have occurred during the calculation or reflect a problem with initial specification of the grid and associated quantities. Running with very small time steps uses up many computational cycles without advancing the solution very far in time. Better to stop and check than to grind for hours.

All of the above take place in a portion of the code known as the pre-processor (Fig. 5.1(a)). The order in which they take place depends on individual code pre-processors, but no calculation will start until all of the above information is specified. The initial conditions and grid are usually displayed in some manner before the calculation proceeds so the analyst can review the problem specification and make corrections if needed. The more user-friendly the pre-processor, the more efficient the use of an analysts' time. Remember that only about 20% of the total cost of an analysis of a particular problem is spent on computing. The remaining 80% involves the analysts' time spent in setting up computational grids and evaluating results. Thus, a powerful post-processor, which can display efficiently the various quantities in the calculation, is a powerful and cost-effective tool. But there is a corollary to this. Powerful, cost-effective, easy-to-use pre- and post-processors also make it easy to *misuse* a code, especially for analysts who lack familiarity with the physics of the problem they are trying to solve.

174   Introduction to Hydrocodes

5.3   Number crunching.

The pre-processor portion of a hydrocode generates the mesh, assigns material characteristics to individual cells or elements, lumps masses, assigns initial and boundary conditions, defines material interfaces and boundaries between the various objects in the calculation and writes this information to a file. Typically this is referred to as a *restart file*. These data will be used to start the calculation at the initial time. The initial time is usually taken to be $t = 0.0$, but this is not a requirement. The starting time can have an arbitrary designation as long as all the information required to run the calculation is available.

As Fig. 5.1(b) indicates, the main processor, or number cruncher, reads the data in the initial restart file and proceeds with the computations indicated. The time is advanced after completion of each loop by an increment δt until the desired problem time is reached. Along the way, one or more restart files are generated. This is done both to save data for later analysis as well as to insure that information is regularly saved in case of system crashes so that long calculations need not be repeated from the beginning.

Appropriately, the coding for these calculations is often contained in a subroutine called LOOP. This may either contain most of the calculations in the loop (Fig. 5.1(b)) as is done in EPIC or serve as the managing routine, calling on subroutines for the components of the calculation, the approach taken in *ZeuS* and DYNA. Let us now look at the components of the loop in turn.

a)   *Velocities and displacements.*

The integration loop begins by solving the equations of motion to determine the accelerations at time t. When a simulation begins, the stress state in the elements is zero (unless a pre-stress has been defined). Thus, nodal accelerations at the beginning of a simulation, $t = 0$, are zero.

The explicit algorithm for determination of velocities and displacements used in all wave propagation codes is very straightforward and economical. An excellent discussion of its advantages and limitations is given in Chapter 1 of Belytschko and Hughes [1983]. The acceleration at time t is used to calculate the velocity and position of a node from the following:

$$\dot{u}\left(t + \frac{\delta t}{2}\right) = \dot{u}\left(t - \frac{\delta t}{2}\right) + (\ddot{u})(t)\delta t, \tag{5.1}$$

$$u(t + \delta t) = u(t) + \dot{u}\left(t + \frac{\delta t}{2}\right)\delta t. \tag{5.2}$$

Note several things:

(1) The above is the central difference algorithm for displacements and velocities from Taylor's theorem. To maintain the central difference nature of the algorithm, it is necessary to compute velocities at half steps and displacements at whole steps. This does not present any computational difficulties.

(2) The central difference algorithm is only *conditionally stable*. If the time step δt is chosen too large the calculation becomes unstable. This occurs very rapidly, in a matter of just a few cycles in Lagrange codes. Belytschko & Hughes [1983] discuss this in some detail and give bounds for the size of the time step.

(3) The time step is not a constant but varies from cycle to cycle. It is determined from the smallest element in the computational mesh. In its simplest form it can be stated as

$$\delta t = k \frac{h_{min}}{c}. \tag{5.3}$$

The h_{min} represents the minimum altitude of all the elements in a calculation and c is the sound speed. Because this derivation is for linear problems, the stability fraction k is attached to allow this to be used as an estimate for nonlinear problems as well. The stability fraction k is taken to be less than 1. Most calculations are done with values of k between 0.6 and 0.9. Keeping $k < 1$ ensures that the time step is always less than the time required to travel across the shortest dimension of the element at the sound velocity of the material. It also ensures that the time step is less than the lowest period of vibration of the system [Belytschko & Hughes, 1983].

While the above points out the salient features of the time step algorithm, in practice, codes such as *ZeuS* and EPIC use the following:

$$\delta t = k \left[\frac{h_{min}}{\sqrt{g^2} + \sqrt{g^2 + c^2}} \right]. \tag{5.4}$$

Here, $g^2 = C_0^2 Q/\rho$, Q is the artificial viscosity; ρ, the density and C_0, the quadratic coefficient of artificial viscosity.

Artificial viscosity has been discussed before. More is available in Belytschko & Hughes [1983] and in Chapter 10 of Zukas *et al.* [1982]. Bulk artificial viscosity is included in codes to be able to treat discontinuities such as shock waves in a continuum calculation. It produces some smearing of the shock intensity locally and allows the shock to be treated as a steep stress gradient rather than a discontinuity, which would require special logic to recognize the formation of shocks and track them throughout

176 *Introduction to Hydrocodes*

the calculation. This is an expensive proposition. Hourglass viscosity is included in some codes (those using quadrilateral and tetrahedral elements) to stabilize the grid.

b)   *Contact.*

Special logic must be incorporated in Lagrange codes to detect contact between objects and to conserve momentum during collisions. These algorithms are known in the literature as slide lines, contact processors, contact–impact algorithms and a few less common names.

In the previous portion of the computational loop, velocities and displacements were updated using velocities and accelerations determined at the previous time step. Recall that at time $t = 0$, the start of the calculation, the accelerations are zero for a stress-free body. For our example of the projectile impact problem described above, we have the situation depicted in Fig. 5.4. This assumes the calculation was started at time

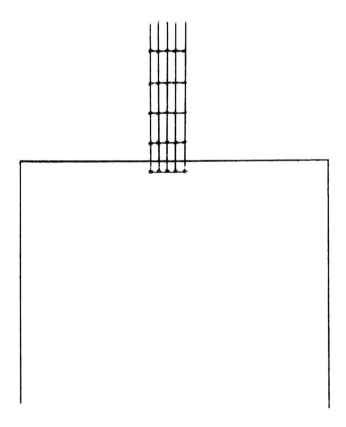

Figure 5.4 Node intrusion on start of calculation.

$t = 0$. Only a few projectile nodes and none of the target nodes are shown for purposes of illustration.

Obviously this is an unacceptable state of affairs as one material cannot occupy the same space as another. The contact algorithm, or slide-line logic, takes care of the intrusion.

In the pre-processor, a series of nodes were defined which make up the master surface. Another set of nodes was defined making up the slave surface. In *ZeuS* this is done automatically for the user. In codes such as EPIC and DYNA, the user must specify master and slave nodes. The designations "master" and "slave" are now mainly historical. In the earliest Lagrange codes, the master surface was checked for intrusion by slave nodes. Thus the designation of master and slave nodes could seriously affect the problem solution. Today most Lagrange codes use a symmetrical treatment. First, intrusion of the master surface by slave nodes is detected. Then the master–slave designations are reversed and the check is made again. This results in much smoother calculations. A very thorough description on numerical modeling of contact processes is given by Laursen [2002].

Once intrusions are detected, the intruding node is moved back onto the master surface and momentum conservation is enforced. Thus, velocities, and subsequent displacements, are imparted to all the objects involved in the calculation.

Once the contact algorithm has moved nodes onto the contact surface and re-computed velocities along that surface, the deformed geometry of each element is computed. These will be used later to determine element stresses and forces.

By way of example, consider the contact algorithm in *ZeuS*. Its general capabilities are described in the manual [(1989)] and other sources [Zukas & Segletes, 1991; Janzon *et al.* 1992]. The contact processor in *ZeuS* is very similar to those in EPIC and DYNA and other Lagrange codes for that matter. It does, however, possess unique features which enhance both user-friendliness as well as computational efficiency.

A unique feature of *ZeuS* is the automatic generation of contact surfaces requiring no user intervention. Contact surface pairs are automatically generated as the mesh is defined. *ZeuS* assumes that any object mesh (a Lagrangian mesh defining all or part of a structure) may come into contact with any other object mesh, and thus create a contact surface between the two objects. When large numbers of object meshes are defined (e.g., a laminated plate, where each lamina is defined with a separate Lagrangian mesh, i.e., each one constitutes an object mesh), the number of contact surfaces automatically defined for n objects will be $n(n-1)/2$. The number of contact surfaces may grow quite large, for even a modest number of object meshes. When certain surface combinations are known, *a priori*, not to undergo contact, these contact surface pairs may be manually removed, in order to gain computational efficiency, but no requirement exists to do so. For example, Figs. 5.5 and 5.6 show the perforation of a 50-ply laminated aluminum plate

178 *Introduction to Hydrocodes*

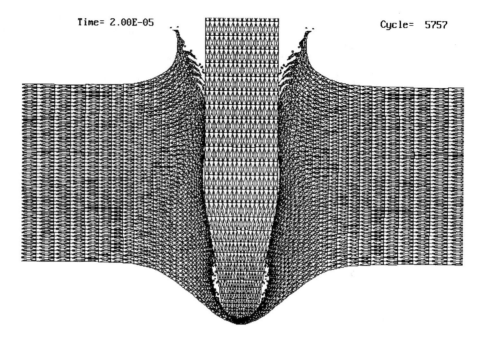

Figure 5.5 Penetration of 50-ply aluminum plate by a steel projectile at 20 μs after impact.

by a steel projectile. Since each plate is known, *a priori*, to contact only its immediate neighbors and the projectile, the initial number of contact surface pairs may be reduced to 99 (49 interplate pairs and 50 projectile-plate pairs), rather than the all-inclusive 1225 contact surface pairs ($= 50 \times 49/2$). Contact surface pairs may even be defined between an object mesh and itself. By doing so, self-intrusion, which may result from excessive bending and straining can be effectively precluded. *ZeuS* makes very limited demand upon computer memory by using indirect addressing to access contact data. It accomplishes this by storing data for all contact surfaces in a single large array, with a secondary pointer array designating the beginning and end of each contact surface. Without the use of such addressing techniques, the alternative is to allocate, for each contact surface, a chunk of memory larger than the largest contact surface, and address it directly. Though such alternatives are simpler to program, they are grossly inefficient in terms of memory allocation. The penalty paid for direct addressing is particularly noticeable on PC architectures, where physical memory is limited. Thus, simply downloading a mainframe code to the PC environment will inevitably place a severe limitation on either problem size, if virtual memory is absent, or on execution speed, if virtual memory is present [e.g., Janzon *et al.*, 1992]. *ZeuS* uses the contact concept discussed above. In this concept, one surface of the contact pair is designated the master

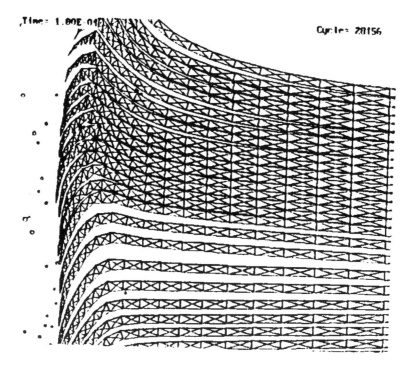

Figure 5.6 Closeup of 50-ply aluminum plate perforation demonstrating both mesh contact-separation and tangential sliding between meshes.

and the other the slave. The search for intrusions of the slave surface into its paired master is performed by sequentially examining each successive element face along the master surface. Upon detection of contact surface intrusion, only the slave surface nodes are relocated back to a point of intrusion upon the master surface. Such a technique introduces an unnatural bias, if special precautions are not taken. In *ZeuS*, two precautions are taken to minimize this bias. First, the designation of master and slave surfaces are reversed at the end of every cycle, so that each member of a contact pair will spend half of its time as a master and half as a slave. Secondly, the direction of search along a master surface is alternated every other cycle, so that any given master surface will be searched in each direction exactly half of the time. Finally, an optional "double pass" contact option is available in *ZeuS*, whereby master and slave surface interchange occurs within the course of a single computational cycle, in order to reduce contact algorithm bias, though at added computational expense. The intrusion search algorithm is further optimized, to take advantage of global (G) object mesh characteristics, to minimize the use of brute force local (L) search techniques. When the contact algorithm is first entered in a given computational cycle, a rectangular box which just encloses each object mesh, is computed. Then, when a given contact surface pair is to be examined for potential contact, a simple global/global (G/G) search is first employed to see if there is any potential contact between the contact object pair. This G/G search entails the trivial determination of whether the rectangular boxes enclosing each object of the contact pair overlap, or not, Figs. 5.7 and 5.8. If not, then no further contact checking need proceed for that contact pair. If so, on the other hand, searches must proceed for each surface element on the master contact surface. In this case, a local/global (L/G) search technique is employed to see whether this individual master element face falls within the rectangular box enclosing the slave object. If not, the search may proceed to the next master element face, Fig. 5.9. Only if this bypass fails, however, does *ZeuS* proceed to the brute force local/local (L/L) search, to ascertain the possibility of one or several individual slaves intruding the designated master element face, Fig. 5.10.

Once intrusion has been detected, an equal and opposite boundary condition must be applied to the opposing contact surfaces, at the point of contact. Momentum is directly exchanged between nodes on the affected surfaces, Figs. 5.11 and 5.12. The method may be approximately thought of as a penalty technique of variable stiffness, the stiffness chosen locally to counter the intrusion in a single integration cycle.

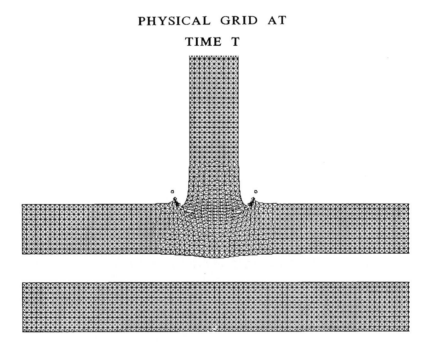

Figure 5.7 Computational grid at time—T.

How Does a Hydrocode Really Work? 181

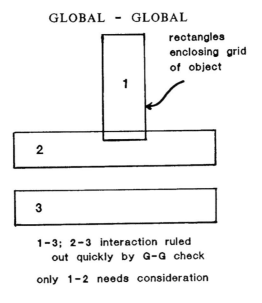

Figure 5.8 Global–global search for intrusion.

In many simulations of violent impact, what was originally a single element mesh may be torn into multiple pieces, as a result of repeated element erosions. The *ZeuS* logic dynamically and automatically creates new objects as well as contact surfaces. Figure 5.13 shows, for example the initial geometry of a 2 cm diameter, $L/D = 0.5$, tool steel slug impacting two 4340 steel plates, at an impact velocity of 2200 m/s. Initially there are

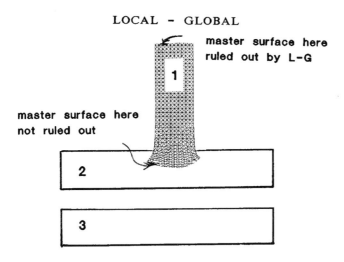

Figure 5.9 Second stage of intrusion search.

182 *Introduction to Hydrocodes*

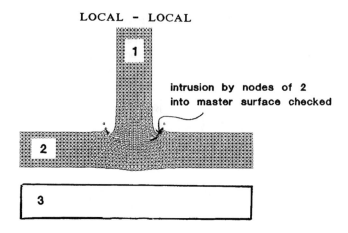

Figure 5.10 Third stage of intrusion search.

three objects and three contact surfaces (1–2, 2–3 and 1–3). As a result of mesh erosion, the simulation at 20 μs consists of 14 object meshes. Figure 5.14 shows the simulation geometry at this time, with the 14 object meshes depicted. Note that, because of the axisymmetric geometry, meshes in the $-R$ half plane are not independent, but are actually mirror reflections of those in the $+R$ half plane. When multiple objects are formed from one, contact surfaces pairs are created between each possible pair of newly created objects. Thus, the 14 object mesh simulation depicted in Fig. 5.14 employs a total of $91 (= 14 \times 13/2)$ contact surface pairs, 88 of which were "grown" from the original set of three pairs. From the point of view of the *ZeuS* erosion processor, a newly created object mesh may consist of a single element (like objects 1,2,3, or 11) or a large element assemblage (like objects 1, 2, 3 or 11). Because a Lagrange mesh deforms with

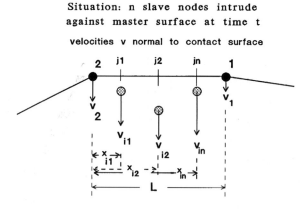

Figure 5.11 Multiple intrusions of slave nodes against master surfaces.

After moving slave nodes onto master surface enforce

Linear Momentum:

$$m_1 v_1 + m_2 v_2 + \Sigma m_i v_i = m_1 v'_1 + m_2 v'_2 + \Sigma m_i v'_i$$

Angular Momentum:

$$m_1 v_1 + 0 + \Sigma m_i a_i v_i = m_1 v'_1 + 0 + \Sigma m_i a_i v'_i$$

where $a_i = x_i / L$

Compatibility:

$$e(v_j - v_i) = (v'_j - v'_i) \qquad \text{(n equations)}$$

[j are points on the master surface corresponding to slave node locations i]

Figure 5.12 Application of momentum conservation and compatibility equations to correct intrusions. The coefficient of restitution is denoted by *e* and ranges from 0 to 1.

the material, very large distortions soon occur in impact problems, which have deleterious effects. Large distortions can be overcome by resorting to rezoning—overlaying a new mesh atop the old one and conserving mass, momentum, energy and the constitutive equation or to erosion. The latter approach is used in *ZeuS* and is very valuable when distortions are highly localized such as in problems involving penetration and perforation. The term "erosion" in this context does not refer to a physical failure mechanism; rather, it describes book-keeping procedures which allow dynamic redefinition of master and slave surfaces when very large distortions occur. The overall result makes it appear that the colliding bodies are "eroding." An example of mesh

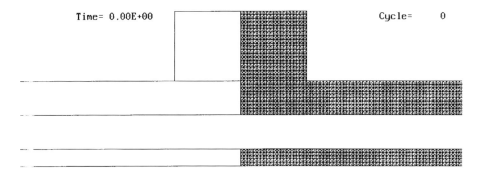

Figure 5.13 Initial geometry of steel slug impact upon two spaced plates: three object meshes, three initial contact pairs.

184 *Introduction to Hydrocodes*

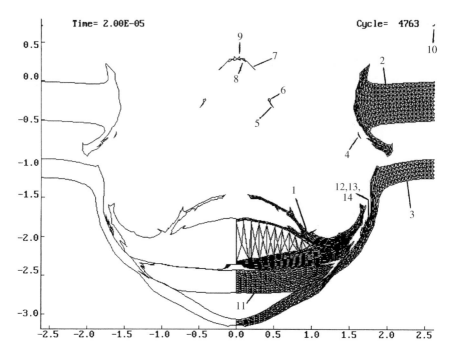

Figure 5.14 Geometry of steel slug impacting two spaced plates at 20 μs after impact. Fourteen (14) current object meshes (shown); 91 current contact surface pairs.

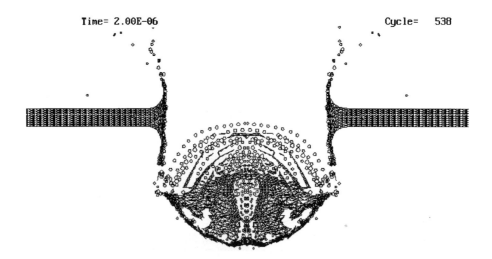

Figure 5.15 Hypervelocity impact of aluminum sphere upon a thin aluminum plate, depicting mesh erosion.

erosion is shown in Fig. 5.15, where the debris cloud resulting from the hypervelocity impact of an aluminum sphere upon a thin aluminum plate is depicted. ZeuS provides the ability to track the material which is separated from the intact Lagrangian mesh as free-flying mass points. These "freeflyers" may interact (via contact) with intact element meshes and, with the appropriate option set, may also interact with each other.

c)   **Strains and strain rates.**

Once nodal positions and velocities have been updated as a result of solving the equations of motion and the actions of the contact processor, we can focus our attention on the elements. Because some element nodes have been moved, deformations may have been introduced into the elements. These lead to stresses, strains and forces which must now be determined.

As was mentioned earlier, in any discrete solution, values of displacements and velocities are determined only at a discrete number of points. To find these and other quantities at other than nodal point, interpolation is required. ZeuS uses *triangular* elements. For these, the velocity at any point in the element may be linearly interpolated from the velocities at the nodal vertices. The interpolation function is

$$\dot{u}_i = \alpha_i + \beta_i r + \gamma_i z, \tag{5.5}$$

where $\dot{u}_i$ refers to the radial (r) and axial (z) velocities (u and v, respectively) within the element, depending on whether $i = 1$ or 2. The parameters α, β, γ are determined from the element geometry and nodal velocities. Strain rates are now computed from

$$\dot{\varepsilon}_r = \frac{\partial \dot{u}}{\partial r} \qquad \dot{\varepsilon}_z = \frac{\partial \dot{v}}{\partial z}, \tag{5.6}$$

$$\dot{\varepsilon}_\theta = \frac{\bar{\dot{u}}}{\bar{r}} \text{ (axisymmetric)} \qquad \dot{\varepsilon}_\theta = 0 \text{ (plane strain)}, \tag{5.7}$$

$$\dot{\gamma}_{rz} = \frac{\partial \dot{u}}{\partial z} + \frac{\partial \dot{v}}{\partial r}; \qquad \dot{\gamma}_{r\theta} = 0; \qquad \dot{\gamma}_{z\theta} = 0. \tag{5.8}$$

Please note several things:

(1) Because of the linear form of the interpolation function for velocities, the strains, and thus strain rates, in an element are constant in space.

(2) Based on the element velocity distribution stated above, the hoop strain rate is not uniform throughout the element. In axisymmetric calculations, this strain rate

component is the only one, which is not constant in space. In practice therefore, an *average* value of hoop strain is used for the element, based on the average radius of the element.

Another important variable is the *effective strain rate*, which is given by

$$\dot{\varepsilon} = \sqrt{[(2/9)[(\dot{\varepsilon}_r - \dot{\varepsilon}_z)^2 + (\dot{\varepsilon}_r - \dot{\varepsilon}_\theta)^2 + (\dot{\varepsilon}_z - \dot{\varepsilon}_\theta)^2 + (3/2)\dot{\varepsilon}_{rz}^2]]}. \tag{5.9}$$

During the subsequent stress calculations, the elastic portion of the total strain rate is calculated and subtracted out to give the equivalent plastic strain rate. This is then integrated and used both as a failure criterion and as a work-hardening parameter.

The volumetric strain and strain rate are required to compute failure. They also appear in the equation of state, which is used to compute very high pressures and they are also needed in the energy equation. The volumetric strain and strain rate are given by

$$\varepsilon_v = \frac{V - V_0}{V_0}, \tag{5.10}$$

$$\dot{\varepsilon}_V = \frac{V^{t+\delta t} - V^t}{\delta t}. \tag{5.11}$$

Finally, it is necessary to account for element rotations. This is handled differently in every code. For the treatment used in *ZeuS*, see pp. 140–141 of the User's Manual and the references therein. For the procedure in other codes, the best source of information is the manual for that code.

Strains are obtained from strain rates by simple integration:

$$\varepsilon_i^{t+\delta t} = \varepsilon_i^t + \dot{\varepsilon}_i^t \delta t. \tag{5.12}$$

d) *Stress, failure and energy for inert materials.*

Once element strains and strain rates have been computed, a hydrocode proceeds to the constitutive relationship to determine the stress state of each element. The simplest of these is the elasto-plastic incremental formulation, which has been incorporated in every code since 1964. It stems from the formulation first used by Wilkins [1964] in HEMP and Johnson [1965] in OIL (the mother of all Euler codes).

We begin by computing the strain rate deviator for isotropic material. If you are a little rusty on this, try the books by Khan & Huang [1995] and Drucker [1967] for a refresher.

$$\dot{e}_r = \dot{\varepsilon}_r - \bar{\dot{\varepsilon}}, \tag{5.13}$$

$$\dot{e}_z = \dot{\varepsilon}_z - \bar{\dot{\varepsilon}}, \tag{5.14}$$

$$\dot{e}_\theta = \dot{\varepsilon}_\theta - \bar{\dot{\varepsilon}}, \tag{5.15}$$

$$\dot{e}_{rz} = \dot{\gamma}_{rz}, \tag{5.16}$$

where

$$\bar{\dot{\varepsilon}} = \frac{(\dot{\varepsilon}_r + \dot{\varepsilon}_z + \dot{\varepsilon}_\theta)}{3}. \tag{5.17}$$

As a result of this definition, we have the fact that $\dot{e}_r + \dot{e}_z + \dot{e}_\theta = 0$. Using this in conjunction with the constitutive equation, the deviatoric stress increments under *elastic* deformation are given by:

$$ds_{rz} = G\dot{e}_{rz}\delta t + (s_r - s_z)\Omega_{rz}\delta t, \tag{5.18}$$

$$ds_z = 2G\dot{e}_z\delta t + 2s_{rz}\Omega_{rz}\delta t, \tag{5.19}$$

$$ds_\theta = 2G\dot{e}_\theta\delta t, \tag{5.20}$$

$$ds_r = 2G\dot{e}_r\delta t - 2s_{rz}\Omega_{rz}\delta t. \tag{5.21}$$

The element rotation Ω is calculated using the Dienes [1979, 1984, 1986] procedure.

So far, the only assumption made is that the material is elastic. The procedure for modeling plasticity is described below. It is known in the literature as the radial return algorithm and is thoroughly described in Belytschko & Hughes [1983]. Many alternatives to this algorithm have been examined, but none have been found to be computationally more efficient than this algorithm first used in HEMP by Wilkins [1964].

To check for material yield, all existing hydrocodes perform the above calculation. The stress deviators are augmented by the above increments. This is a completely elastic calculation. Next, the von Mises yield function is computed from

$$f = (1/\bar{Y})\sqrt{\left[\frac{3}{2}(s_r^2 + s_z^2 + s_\theta^2) + 3s_{rz}^2\right]}. \tag{5.22}$$

Here, $\bar{Y}$ is the dynamic flow stress and f is the yield function. If $f > 1$, yielding has occurred.

This apparently simple model has considerable flexibility. The dynamic flow stress may be modeled with dependence on many factors such as pressure, temperature, effective plastic strain and strain rate. A common formulation is

$$\bar{Y} = Y(\varepsilon_p)(1 + C_1 \ln(1 + d\varepsilon/dt))(1 + C_2 p + C_3 p^2)(1 + C_4 + C_5 \theta + C_6 \theta^2). \tag{5.23}$$

The Cs are empirical constants; p, the hydrodynamic pressure; θ, the absolute temperature (degrees Kelvin); $d\varepsilon/dt$, the rate of change of equivalent strain with respect to time; $Y(\varepsilon_p)$, the nominal flow stress based only on work-hardening effects and $\bar{Y}$, the dynamic flow stress.

The nominal value of work-hardened flow stress $Y(\varepsilon_p)$ is made to vary linearly between the yield stress Y_0 and the ultimate stress Y_{ult} for equivalent plastic strains between zero and the ultimate strain ε_{ult}. This relationship demonstrates the work-hardening effect and is given as

$$Y(\varepsilon_p) = Y_0 + (Y_{\text{ult}} - Y_0)(\varepsilon_p/\varepsilon_{\text{ult}}) \quad \text{if } \varepsilon_p < \varepsilon_{\text{ult}}, \tag{5.24}$$

$$Y(\varepsilon_p) = Y_{\text{ult}} \text{ if } \varepsilon_p \geq \varepsilon_{\text{ult}}. \tag{5.25}$$

This again proves the old adage that there is no free lunch. Having added some flexibility to our material model, we now find the need to define two additional parameters, the pressure and temperature, and come up with 6 (count them, 6) empirical constants. Needless to say such detailed information is hardly ever available in the literature. Hence, you must go off to the lab and come up with six constants, ever mindful of Cauchy's remark that with four free parameters he could derive an equation for an elephant and with five he could make its tail wiggle!

Cope for a bit with this daily nightmare of code users (some have been known to wake up in the middle of the night in a cold sweat screaming "But where do I get the data?"). It is a humbling experience and partly explains why the divorce rate among computing professionals is above the national average. (The other reason is that they spend more time with their computers than their families. Hence many satisfy the definition of a theoretical physicist: someone whose existence is postulated to satisfy

Conservation of Mass but who is actually never seen in reality.) In the meantime, we will work on the other two parameters.

By the way, while you are working on those six constants, be sure that they are structured in such a way that equating all the constants to zero will make the material exhibit no dependence on strain rate, pressure or temperature.

Absolute temperature is computed from

$$\theta = \theta_0 + \frac{E_s}{C_s \rho_0}, \tag{5.26}$$

where θ_0 is the initial absolute temperature; E_s, the internal energy per unit original volume; C_s, the specific heat; and ρ_0, the initial density of the material. Did you notice how subtly another parameter, E_s, crept into the analysis?

How long since you have worked with absolute temperature? Remember degrees Kelvin? Check your first year physics text if you need a refresher.

C_s is the specific heat. Which one? There are two of them. If you cannot answer in 60 s, go back to your introductory thermodynamics book.

Coming up with pressure is just a wee bit more involved. In the elastic region, pressure is computed from the sum of the normal stresses

$$p = \frac{\sigma_{11} + \sigma_{22} + \sigma_{33}}{3}. \tag{5.27}$$

Once the elastic range is exceeded, pressure is determined from an equation of state. Several examples were given earlier. There is an absolutely outstanding description of equations of state for solids in Drumheller [1998] and for explosives in Mader [1979, 1997]. Most codes incorporate the Mie–Gruneisen equation of state. For hypervelocity impacts, many Euler codes use the Tillotson EOS or some variant thereof. CTH uses the SESAME tables, a table look-up form of an EOS that requires interpolation between tabulated values.

The Mie–Gruneisen form, which is used in ZeuS, EPIC and many other codes incorporates a polynomial fit to Hugoniot data and an energy term to get off the Hugoniot. It looks like this:

$$p = (K_1\mu + K_2\mu^2 + K_3\mu^3)\left(1 - \frac{\Gamma\mu}{2}\right) + \Gamma(I - (1+\mu)I_0) \text{ if } \mu \geq 0, \tag{5.28}$$

$$p = (K_1\mu)\left(1 - \frac{\Gamma\mu}{2}\right) + \Gamma(I - (1+\mu)I_0) \text{ if } \mu < 0. \tag{5.29}$$

In the above p is the hydrodynamic pressure; μ, the compression, expressed in terms of initial and present volumes as $(V_0 - V)/V$; $K_i(i = 1, 3)$ are equation of state parameters from tabulations or experiments; Γ, the Gruneisen coefficient; I, the internal energy per

190 Introduction to Hydrocodes

unit current volume; and I_0, the original internal energy per unit initial volume (ambient conditions).

Several important points regarding the equations of state should be remembered:

(1) The Mie–Gruneisen form assumes no phase changes. Hence, if a body starts as a solid, it remains a solid if you use Mie–Gruneisen. If, in actuality, with increasing temperature and pressure the materials goes from a solid, to liquid to gaseous phase, then a different EOS such as Tillotson, Gray, Los Alamos, etc. must be used.

(2) The Gruneisen parameter is a function of volume. However, most codes, *ZeuS* included, use it as a material parameter and treat it as a constant. Under states of large compression, retention of the Gruneisen gamma as a constant will cause the Mie–Gruneisen equation to act in a non-physical way—additional compression will cause the pressure to decrease and the sound speed to go negative.

(3) Equations of state are derived from shock-loading experiments and thus are really intended for use at high pressures. The cubic fit through Hugoniot data will produce good results for pressure under loading conditions where pressure exceeds material strength by a large margin. It is likely to give poor results in regions where pressure and stress are of the same order of magnitude. Good results for elastic and elasto-plastic wave propagation problems have been obtained with hydrocodes, but care must be taken to remove the energy coupling for low velocity impacts and insure that the cubic fit goes through a state of zero pressure for zero compression. For elastic calculations it is best to obtain pressure as the sum of stress states or, if an EOS is used, to specify only the K_1 parameter and set that to the bulk modulus.

Because the Mie–Gruneisen equation of state couples internal energy into the equation of state, the energy equation and the equation of state must be solved simultaneously.

The energy equation solved with the EOS in the code is a discretized version of

$$E^{t+\delta t} - E^t = \int_V \int_t^{t+\delta t} (\sigma_i - Q_i)\dot{\varepsilon}_i \mathrm{d}t/\mathrm{d}V. \tag{5.30}$$

Here, E represents internal energy per unit current volume at the times indicated, V represents the element current volume, σ_i and $\dot{\varepsilon}_i$ are components of the stress vector and the strain rate vector and Q is the artificial viscosity. Recall that the artificial viscosity is in codes for two reasons. Bulk viscosity is there to smear shocks and allow what amounts to mathematical discontinuities to be treated in a continuum fashion. Hourglass viscosity is included for certain element types (quadrilaterals and hexahedral) to prevent non-physical distortions.

The current sound speed in an element is used in determination of the time step for each cycle in the loop. The speed of sound in a material is given by $c^2 = \partial p/\partial \rho$.

The exact form will depend upon the equation of state being used. For Mie–Gruneisen, the form is

$$c^2 = \frac{1}{\rho_0}[K_1(1 - \Gamma\mu) + K_2(2\mu - 1.5\Gamma\mu^2) + K_3(3\mu^2 - 2\Gamma\mu^3) + \Gamma I(1 + \mu) \\ + \Gamma PI(1 + \mu)], \tag{5.31}$$

where c is the speed of sound in the inert material; ρ_0, the initial element density; K_i ($i = 1, 3$), the Mie–Gruneisen EOS parameters; Γ, the Gruneisen coefficient; $\mu[= (V_0 - V)/V]$, a measure of the element compression; I, the internal energy per unit current volume; P, the pressure which has been augmented by the value of the artificial viscosity.

The above diversion was required to allow us to compute element plasticity. The procedure is known in the literature as the *radial return algorithm* and is used in all hydrocodes. Belytschko & Hughes [1983] has a detailed description of the algorithm as well as alternative methods for computing plastic response.

Recall that all hydrocodes assume *a priori* that material behavior is elastic. Once the elastic calculations are completed, a check is made for each element to see if the von Mises yield criterion has been violated. If it is determined that yielding has occurred, several tasks are performed:

(1) The assumed stress state is updated to reflect the occurrence of plastic flow.

(2) The equivalent plastic strain is augmented by the amount of plastic strain occurring that cycle.

(3) The element is tested for failure.

The rules of plasticity dictate that the stress state of any material must lie within (or on) the yield surface defined by the yield criterion. Energy conditions further dictate that the plastic strain increment be normal to the yield surface. Because the normal to the von Mises yield surface never contains a hydrostatic component, the plastic strain increment contains no hydrostatic component. Furthermore, for isotropic materials obeying the von Mises condition, the vector normal to the yield surface is proportional to the stress deviator state. Thus, a good first-order approximation to the stress state that should exist at the end of a computational cycle may be obtained by scaling back each component of the *a priori* assumed stress state uniformly until the yield surface is reached.

During periods of plastic flow, the equivalent plastic strain is updated. Knowing the stress state at the beginning of the time cycle and the updated stress state which was acquired via stress scale-back, an equivalent elastic strain increment is computed for the element. This elastic strain increment is subtracted from the total strain increment to give the equivalent plastic strain increment. The equivalent plastic strain for an element is determined by keeping track of all the plastic strain increments.

Having the complete stress and strain state for each element, we can now determine whether material failure has occurred. The various failure criteria used in hydrocodes have been reviewed by Zukas [1987]. *ZeuS* allows two forms of material failure: shear failure and total failure. A material, which fails in shear, can carry only compressive stresses (no shear or tension). This occurs when the *equivalent plastic strain* exceeds a preset value. The other type is total failure. Whereas shear failure crudely represents physical failure, total failure is used when it is no longer computationally efficient to retain an element in the mesh. Total failure is accompanied by erosion of the material from the element mesh and can occur for a variety of reasons. The primary ones are that the *equivalent plastic strain* of an element has exceeded a preset value or the *volumetric strain* has exceeded an allowable preset value. Other conditions arise on rare occasions, which lead to total failure. These are usually associated with sudden changes in the element state which, in turn, are usually a result of contact surface nodal relocation.

Though element removal (erosion) associated with total element failure has the appearance of physical material erosion, it is, in fact, a numerical technique used to permit extension of the computation. Without numerical erosion, severely crushed elements in Lagrangian calculations would drive the simulation time step to a very small value, resulting in the expenditure of very many computational cycles with negligible advance in the simulation time. Also, Lagrangian elements which have become very distorted have a tendency to "lock up," thereby inducing unrealistic distortions in the computational mesh. If the erosion strain is set too low, there is unrealistic excessive removal of material. If it is set too high, grid lockup and minimum time step violations can occur. Convergence of results generally occurs for plastic strain values between 1.5 and 2.5.

When total failure occurs, *ZeuS* erodes the element from the mesh. This requires re-evaluation of the object to determine its new free surface. If necessary, *ZeuS* creates new free objects (child meshes) when erosion causes an object mesh (parent mesh) to break into multiple fragments. Each child mesh may interact will all other meshes in the calculation.

When a node finds itself no longer attached to an unfailed element, the node becomes a free-flying node. A free flyer becomes in essence a surface of its own and may interact with other meshes and other free flyers in a calculation.

Existing failure models in all codes assume that material failure is an instantaneous process. When the maximum of a particular field variable is reached, be it some form of stress, strain or energy, failure occurs immediately. This assumption works well for energetic impacts, such as hypervelocity impact, where there is little time for micromechanical processes to work. The lower the energy imparted to the material, the worse the assumption becomes. Dynamic fracture is generally taken to be a four-part process [Shockey, Seaman & Curran, 1973; Shockey, Curran & Seaman, 1978]:

- rapid nucleation and microfractures at a large number of locations in the material;
- growth of the fracture nuclei in a rather symmetric manner;
- coalescence of adjacent microfractures;
- spallation or fragmentation by formation of one or more continuous fracture surfaces throughout the material.

A large number of experiments, dealing primarily with spallation, have indicated that a time-dependence for failure must be incorporated into a failure model if it is to have a reasonable hope to be successful. Hence, the weakest portion of any hydrocode is its treatment of material failure. The major reasons why calculations fail to agree with observation are: (a) human error; (b) use of inappropriate material data for the selected material models, e.g., use of static data for dynamic calculations at high strain rate; and (c) use of inappropriate or incomplete models for the physics of the situation. The first two can be overcome. The third still requires research to improve our knowledge of material behavior, especially failure, at high strain rates.

e)   ***Pressure and energy for explosive materials.***

Explosive materials are treated by wave propagation codes in an inherently different manner than inert materials. First of all, explosive materials are not able to support shear stresses, so that only pressure need be modeled. Secondly, there is no concept of material yield for explosives. Finally, the equation of state used to calculate pressure in explosive materials is different from that used in inert materials.

The gamma law EOS is used to calculate pressure for explosive materials. It is the one most commonly found in production wave propagation codes, but there are a multitude of alternatives, e.g., see Davis [1998] and Mader [1979, 1997]. The form used by *ZeuS*, HEMP [Giroux, 1973], EPIC [Johnson, 1978] and other Lagrange codes is given by

$$P = F(\gamma - 1)I, \qquad (5.32)$$

where P is the detonation pressure; F, the burn fraction of the material element; γ, a material constant; and I, the internal energy per unit current volume.

The burn fraction F is a parameter which determines the percent of an explosive element which has been explosively reacted. It will equal zero initially. It can rise above zero in one of two ways:

(a) When the wave traveling at the detonation velocity reaches the element. The statement for this criterion is

$$F_1 = \frac{(t - t_s)D}{4I(V/V_0)}, \tag{5.33}$$

where D is the detonation velocity; t, the current time; t_s, the time required for the wave traveling at the detonation velocity to reach the node of the element closest to the detonation point; I, the internal energy per unit current volume; V, the current element volume; and V_0, the initial element volume.

(b) When the element becomes compressed through the motion of adjacent elements. This criterion permits a converging detonation wave to travel at a speed greater than the detonation velocity since it is based on element compression. It is given by

$$F_2 = [1 - (V/V_0)](1 + \gamma). \tag{5.34}$$

The burn fraction used is the maximum of F_1 or F_2. As stated previously, the minimum value for $F = 0$. Furthermore, the maximum value for $F = 1$.

Like the Mie–Gruneisen equation, energy and pressure equations must be solved simultaneously, with artificial viscosity augmenting the computed value of pressure. Again, a subcycling scheme is used to minimize on error in the pressure–energy calculation. The speed of sound in explosive materials (which is used to compute artificial viscosity and integration time step) is computed from a different relationship for explosives. The form is given by

$$c^2 = \frac{\gamma P}{\rho}, \tag{5.35}$$

where c is the speed of sound in an explosive element; γ, a constant parameter that characterizes the explosive; P, the element pressure which has been augmented with artificial viscosity; and ρ, the element density.

This approach to energetic materials is adequate if the sole interest in the explosive is the amount of energy it can transfer to a surrounding structure, be it a shaped charge, a mining or blasting operation or loading a nearby structure. More sophisticated models must be used to study fundamental aspects of explosive behavior such as initiation and propagation of a detonation front; see Mader [1979, 1997], Zukas & Walters [1998] and Cooper [1996] for details.

f)   *Nodal forces.*

Why have we gone through all this effort? We have done it to be able to do the last step in the process, compute $F = ma$. Once we have the accelerations at the current cycle, we can update the velocities and displacements for the next time step and go through the loop as many times as necessary to solve our problem.

We now have everything we need. We have updated element geometries, we have stresses and strains—in short, all the ingredients to compute forces.

This is done in different ways in different codes. In EPIC [Johnson, 1978], the concentrated forces which act on the concentrated masses at the nodes are obtained by calculating forces that are statically equivalent to distributed stresses in the elements. Referring to Fig. 5.16, the radial, axial and tangential forces acting on node i of an element are

$$F_r^I = -\pi\bar{r}[(z_j - z_m)\sigma_r + (r_m - r_j)\tau_{rz}] - \frac{2\pi A \sigma_\theta}{3}, \tag{5.36}$$

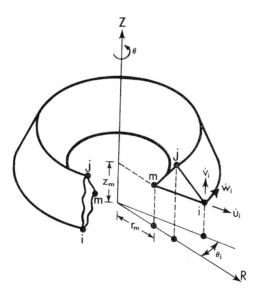

Figure 5.16 Triangular cross-section element geometry.

$$F_z^I = -\pi\bar{r}[(r_m - r_j)\sigma_z + (z_j - z_m)\tau_{rz}], \tag{5.37}$$

$$F_\theta^I = -\pi\bar{r}\left[\frac{\bar{r}}{r_i}(z_j - z_m)\tau_{r\theta} + (r_m - r_j)\tau_{z\theta}\right]. \tag{5.38}$$

In *ZeuS* a contour integration is used to obtain concentrated nodal forces. This procedure helps to reduce difficulties near the centerline.

After element properties have been characterized through the stress and strain calculations, it is time to shift attention back to the nodes. In particular, this last stage of integration loop uses the element stresses to evaluate equivalent concentrated forces at the nodes. These forces are used to solve the nodal equations of motion at the beginning of the subsequent integration cycle. Every node that is part of an intact mesh lies at the vertex of at least one element. In general though, a node may be a vertex of an arbitrary number of elements. Consider the node displayed in Fig. 5.17 which is a vertex for six elements in addition to lying on a free surface. If a control volume is created around the node as shown, the element stresses may be integrated along the control volume surface to yield an equivalent concentrated nodal force. There are several key features of the control volume construction about a node. They will be described below in terms of the control contour, which is defined as the intersection of the control surface with the *R–Z* plane: (1) the control contour forms a series of line segments; (2) the control contour cuts through the geometric center of every attached element; (3) the control contour cuts through the midpoint of all

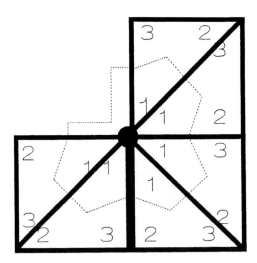

Figure 5.17 Contour integration for force calculation.

element edges on which the node lies; and (4) when the control contour hits a free surface of the mesh, it follows a path parallel to the free surface, located an infinitesimal distance from the free surface, towards the node in question. This construction is such that any segment of the control contour within an element is also a segment on a control contour of an adjacent node, except that the outward normals are of opposite sense. Similarly, the area located an infinitesimal distance from the free surface of a mesh is stress free, so that there are no concentrated force contributions from the free surface control contours. As such, every equivalent concentrated force contribution on the node is balanced by equal and opposite contribution on adjacent nodes, in accordance with Newton's laws. As shown in Fig. 5.17 if one assumes the node in question to have the arbitrary designation of "1" for each of the n attached elements, with nodes "2" and "3" of each element progressing counterclockwise around the element in the R–Z plane, then for the case of plane strain, the concentrated force per unit thickness on the node in question works out to be

$$F_r = \sum_1^n -(1/2)[(z_2 - z_3)\sigma_r + (r_3 - r_2)\sigma_{rz}], \tag{5.39}$$

$$F_z = \sum_1^n -(1/2)[(r_3 - r_2)\sigma_z + (z_2 - z_3)\sigma_{rz}], \tag{5.40}$$

where r_i and z_i refer to the R and Z coordinates of node i for each node participating in the summation.

These equations are incidentally equivalent to the forms presented by other codes such as EPIC for the concentrated force calculation. However, when the contour surface stress integral is performed for the case of axisymmetry, a form of the concentrated force equations arises, which is similar but not identical to the forms proposed by authors of other codes. Using the same node numbering convention described above, the concentrated force on the node is given by:

$$F_r = \sum_1^n -\pi\{[(\overline{z_{12}} - \bar{z})(\overline{r_{12}} + \bar{r}) + (\bar{z} - \overline{z_{13}})(\bar{r} + \overline{r_{13}})]\sigma_r + [(\overline{r_{13}} - \bar{r})(\overline{r_{13}} + \bar{r}) \\ + (\bar{r} - \overline{r_{12}})(\bar{r} + \overline{r_{12}})]\sigma_{rz} + (2/3)A\sigma_\theta\}, \tag{5.41}$$

$$F_z = \sum_1^n -\pi\{[(\overline{r_{13}} - \bar{r})(\overline{r_{13}} + \bar{r}) + (\bar{r} - \overline{r_{12}})(\bar{r} + \overline{r_{12}})]\sigma_z + [(\overline{z_{12}} - \bar{z})(\overline{r_{12}} + \bar{r}) \\ + (\bar{z} - \overline{z_{13}})(\bar{r} + \overline{r_{13}})]\sigma_{rz}\}, \tag{5.42}$$

where

$$\bar{r}_{ij} = (1/2)(r_i + r_j), \tag{5.43}$$

$$\bar{r} = (1/3)(r_1 + r_2 + r_3). \tag{5.44}$$

The $\bar{z}_{ij}$ and $\bar{z}$ are calculated in a similar fashion to $\bar{r}_{ij}$ and $\bar{r}$ and A is the element area.

These axisymmetric concentrated force equations will reduce to those used in EPIC if one assumes in the current equations that $\bar{r}_{ij} = \bar{r}$. The error introduced by making this assumption becomes smaller the further an element is from the axis of symmetry, but is significant for elements adjacent to the axis of symmetry.

5.4 Once more, with feeling.

We have come to the end of our journey for a single time step δt. The next step is to integrate the equations of motion to obtain

$$\ddot{u}_i = \frac{\bar{F}_r^i}{m_i}. \tag{5.45}$$

Then, as we did at the beginning of this section, the velocities and displacements for the next time increment are obtained and the procedure continued until the desired solution time is reached.

REFERENCES

ZeuS Technical Description and User's Manual (1987–1998), Baltimore, MD: Computational Mechanics Consultants, Inc.

T. Belytschko; T.J.R. Hughes (eds). (1983). *Computational Methods for Transient Analysis*. Amsterdam: North-Holland.

P.W. Cooper. (1996). *Explosives Engineering*. New York: Wiley-VCH.

W.C. Davis. (1998). Shock waves; rarefaction waves; equations of state. In J.A. Zukas and W.P. Walters, eds, *Explosive Effects and Applications*. New York: Springer-Verlag.

J.K. Dienes. (1979). On the analysis of rotation and stress rate in deforming bodies. *Acta Mech.*, vol. 32, 217.

J.K. Dienes. (1984). The effect of finite rotation on a problem in plastic deformation. In A.S. Khan, ed, *Proceedings of the International Symposium on Plasticity*, Bell Anniversary Volume, Norman, OK, July 30–August 3.

J.K. Dienes. (1986). A discussion of material rotation and stress rate. *Acta Mech.*, vol. 65, 1-11.

D.S. Drumheller. (1998). *Introduction to Wave Propagation in Nonlinear Fluids and Solids*. Cambridge, U.K.: Cambridge U. Press.

D.C. Drucker. (1967). *Introduction to Mechanics of Deformable Solids*. New York: McGraw-Hill.

E.D. Giroux. (1973). *HEMP User's Manual*, Lawrence Livermore National Laboratory, Report UCRL-51079, Rev. 1.

B. Janzon, N. Burman, J. Forss, S. Karlsson and E. Liden. (1992). EFP modeling by numerical continuum dynamics on personal computers – a comparison between PCDyna2D, ZeuS and Autodyn-2D, *Proc. 13th Intl. Symp. On Ballistics*, Stockholm, Sweden.

W.E. Johnson. (1965). *OIL, a continuous two-dimensional Eulerian hydrodynamic code*, General Atomic Corp., Report GAMD-5580, San Diego.

G.R. Johnson. (1978). EPIC-2, a computer program for elastic–plastic impact computations in 2 dimensions plus spin, U.S. Army Ballistic Research Laboratory, Report ARBRL-CR-00373, June 1978.

A.S. Khan and S. Huang. (1995). *Continuum Theory of Plasticity*. New York: Wiley.

T.A. Laursen. (2002). *Computational Contact and Impact Mechanics: Fundamentals of Modeling Interfacial Phenomena in Nonlinear Finite Element Analysis*. New York: Springer-Verlag.

C.L. Mader. (1979). *Numerical Modelling of Detonation*. Berkeley: U. of California Press.

C.L. Mader. (1997). *Numerical Modelling of Explosives and Propellants*. Boca Raton: CRC Press.

D.A. Shockey, L. Seaman and D.R. Curran. (1973). The influence of microstructural features on dynamic fracture. In R.W. Rhode, B.M. Butcher, J.R. Holland and C.H. Karnes, eds, *Metallurgical Effects at High Strain Rates*. New York: Plenum.

D.A. Shockey, D.R. Curran and L. Seaman. (1978). Computer modeling of microscopic failure processes under dynamic loads. In K. Kawata and J. Shioiri, eds, *High Velocity Deformation of Solids*. New York: Springer-Verlag.

M.L. Wilkins. (1964). Calculation of elastic–plastic flow. In B. Adler, *et al.*, eds, *Methods in Computational Physics*. New York: Academic Press.

J.A. Zukas. (1987). In T.Z. Blazynski, ed, *Materials at High Strain Rates*. London: Elsevier Applied Science.

J.A. Zukas, T. Nicholas, H.F. Swift, L.B. Greszczuk and D.R. Curran. (1982). *Impact Dynamics*, New York: Wiley, reprinted 1992 by Krieger Publishing Co., Melbourne, FL.

J.A. Zukas and S.B. Segletes. (1991). Hypervelocity impact on space structures. In T.L. Geers and Y.S. Shin, eds, *Dynamic Response of Structures to High-Energy Excitations*. PVP-Vol. 225, New York: ASME.

J.A. Zukas and W.P. Walters (eds.) (1998). *Explosive Effects and Applications*. New York: Springer-Verlag.

Chapter 6
Alternatives to Purely Lagrangian Computations

6.1 Introduction.

Thus far, we have been looking at Lagrangian techniques for dealing with problems involving fast, transient loading. Lagrangian methods offer several advantages [Herrmann, Bertholf & Thompson, 1974] over the competition. But because Lagrangian codes cannot solve all the problems involving fast short-duration loading, honorable mention must be given to other techniques. The more popular alternative methods, Euler codes, coupled Euler–Lagrange codes, arbitrary Lagrange–Euler (ALE) techniques and Meshless Methods are described briefly in this chapter.

If we compare Lagrange methods with the competition, we find that:

- The governing equations (conservation of mass, momentum and energy) are simpler because of the absence of convective terms representing mass flow in the coordinate frame. As you will see directly, the general conservation equations for a continuum allow for the variation of the state of a material in both space and time. The motion of material from one cell to another (finite differences) or one element to another (finite elements) is referred to as *material transport*. In a Lagrange code, the mass of a cell or element at the beginning of a computation remains the same throughout the computation. The grid distorts with the material so density and volume change but mass remains constant. Thus, conservation of mass is automatically satisfied. By contrast, the computational grid is fixed in an Euler code so that material flows through it in all coordinate directions. Here the volume of a computational cell or element is fixed while mass and density vary. Thus, at any given time step in any given cell/element, not only do we need to worry about the time rate of change of material variables but their spatial distribution as well.

- Because the grid distorts with the material in Lagrange computations, material interfaces and surfaces are stationary in the material coordinate frame. They are sharply defined at each increment of time and interface conditions such as contact as well as boundary conditions can be treated in a straightforward manner. Recall that in a Lagrange code, the critical component was the contact processor. The bulk of the computational time is spent in the routines defining material interface conditions. The analogous situation in Euler codes is accounting for material transport. Because the grid is fixed and material flows through it, interfaces are best known to within one mesh dimension. Complex logic is required to specify material interfaces and material boundaries with greater precision. More on this later.

- Material histories required in some constitutive equations are easily accounted for in Lagrange calculations. But because of material diffusion, material histories are obtained with great difficulty on Euler codes, if at all. Material in any given cell at a given time will be moved and may be replaced with material from several different cells at the next time increment. Complex logic may be required. An example is the incorporation of nucleation and growth fracture models in the HELP code [Hageman & Herrmann, 1977] which required a parallel set of Lagrange calculations to track the evolution of damage within the Eulerian framework.

The advantages of Lagrange codes are offset by grid distortion. With large distortions the time increment for advancing the computations is forced to approach zero thus rendering the calculations uneconomical. The use of sliding interfaces and rezoning can extend the range of applicability of Lagrange codes to larger distortions. With the advent of eroding slide lines, as discussed in the previous chapter, some Lagrange calculations (those involving primarily highly localized distortions such as would be encountered in penetration of projectiles into thin or thick targets at velocities up to the hypervelocity range) can now perform calculations that previously had been exclusively in the domain of Euler codes (e.g., see Zukas, Segletes & Furlong [1991], Zukas, Furlong & Segletes [1992], and Zukas & Segletes [1991]). Examples are shown in Figs. 6.1 and 6.2.

Similarly, the ability to handle large distortions in Euler codes is offset by the need to account for material transport. This is a non-trivial and persistent problem, affecting not only the accuracy of the calculation but its economics as well. Calculations taking up to 100 h on the fastest available computers are a luxury in a research environment and an impossibility in a practical setting. Experimental facilities for impact studies can take a few weeks to a month to set up for a particular project. Once in place, however, one to four experiments (data points) per day can be generated, depending on the scale of the experiment. Things go much faster in enclosed indoor facilities with set instrumentation than they do if the experiments are done in the open using large-caliber launchers and cranes to move portions of the experiment into position. Aside from the additional work and safety considerations, there are the vagaries of weather. Working in rain, snow, mud and extremes of temperature is not all that enjoyable, and then there are the various ways that lightning finds to interact with the miles of wiring laid out for outdoor tests.

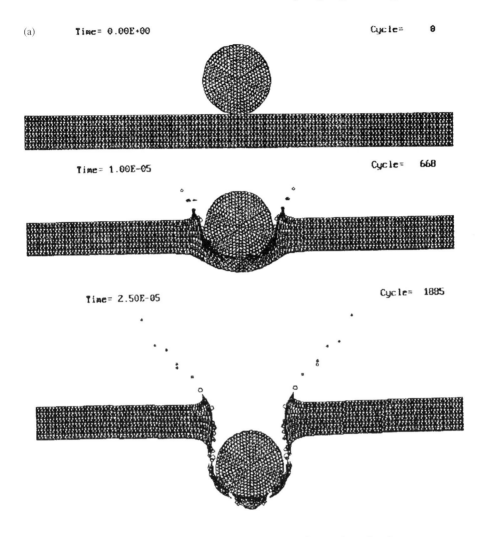

Figure 6.1 (a) *ZeuS* code calculation of steel sphere impacting aluminum target at 800 m/s: deformation patterns at various times after impact.)

In short, this is another one of Mother Nature's ways of telling us that there is no free lunch. The choice between Lagrange and Euler techniques for a given application is rarely clear-cut. Rather, it is a matter of tradeoffs involving accuracy, economics and time constraints. Sometimes it is better not to compute. If your problem comes with a finite budget and must be completed in a finite time, the use of experiments and guesstimates based on simple models will get you further than reliance on a calculation that exceeds your available resources by several orders of magnitude.

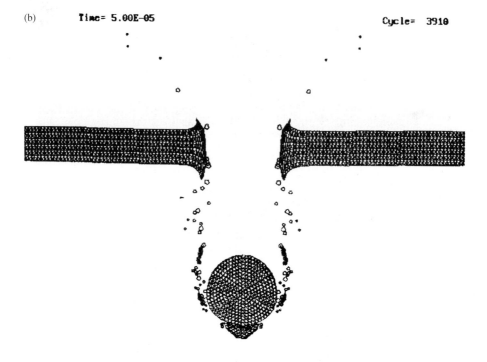

Figure 6.1 (b) *ZeuS* code calculation of steel sphere impacting aluminum plate at 800 m/s: target perforation.

6.2   Euler codes.

Pure Euler techniques have been around for the better part of a century. They are ideal for handling large distortions. Accurate treatment of surface motions, material interfaces—boundaries in general—is another matter. Of course, this latter problem poses no difficulties for Lagrange techniques. However, severe distortions of all or part of a Lagrange grid can bring the calculation to a halt before the phenomena of interest can be computed. To overcome these problems, development began in the mid-1950s to develop Eulerian techniques that could accurately treat surface motion, for application primarily to problems involving hypervelocity (striking velocities in excess of 6 km/s). The earliest examples of such codes were OIL, developed by Wallace Johnson, and HELP, developed by J.M. Walsh and his associates. The codes were not totally independent of each other. Their developers were a small, elite group of physicists and mathematicians who, like

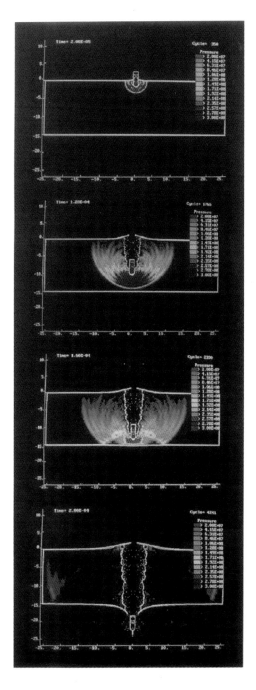

Figure 6.2 *ZeuS* code simulation of hydrodynamic ram event [Zukas *et al.*, 1992].

wandering gypsies, tended to follow the ebb and flow of government contracts, sometimes working together at one laboratory, sometimes separately, depending on funding. Thus, HELP [Hageman & Walsh, 1971] evolved from four hydrodynamic codes over a span of 15 years. HELP was an extension of the one-material code RPM [Dienes, Evans, Hageman, Johnson & Walsh, 1968] which itself was an extension of OIL [Johnson, 1965]. OIL in turn evolved from PIC [Harlow & Evans, 1957] by replacing the discrete particle formulation in PIC with a continuous material description. For those fascinated by the history of hydrocodes, I recommend the articles by Johnson tracking the efforts in computational fluid dynamics at Los Alamos from the PIC codes to the present [Johnson, 1996] and the history of hydrocodes applied to hypervelocity problems by Johnson and Anderson [Johnson & Anderson, 1987].

So where does one start when developing an Euler code? Well, when in doubt, the best place to start is at the beginning. In this case it is the conservation equations of continuum mechanics. If you are a touch rusty on the subject I recommend enthusiastically the delightfully lucid treatment in Anderson [1995]. Other good sources would be the books by Fung [1965], McDonald [1996], Mase & Mase [1999], Faber [1995], Shyy, Udaykumar, Rao & Smith [1996] or any of the many books available on continuum mechanics. To illustrate the approach we follow the development in HELP [Hageman & Walsh, 1971].

Euler codes divide space into fixed cells through which material moves. Because the initial codes were designed to solve problems involving hypervelocity impact, where pressures generated on impact were orders of magnitude larger than material strength, the material was thought of as a fluid. The problem with this approach was that the early stages of impact and crater formation were modeled quite well. However, the codes lacked a mechanism (such as material strength) to prevent the crater growth from being arrested. Natural viscosity proved too weak a mechanism (not to mention the fact that viscosities for solids such as steel and aluminum are not readily found in handbooks). Hence, once a crater formed and started to grow it never stopped. To remedy this, material strength in the form of a simple incremental rigid plastic or elastic-perfectly plastic formulation was added to the codes. The incorporation of such a model in HELP was what made it a novelty and a very popular code in its time.

The equations of motion (conservation equations) of the rate of change of mass, momentum and energy in an Eulerian cell may be written as [Hageman & Walsh, 1971]:

$$\frac{\partial \rho}{\partial t} = -\frac{\partial(\rho u_i)}{\partial x_i}, \tag{6.1}$$

$$\rho \frac{Du_j}{Dt} = \frac{\partial}{\partial x_i}(\sigma_{ij}), \tag{6.2}$$

$$\rho \frac{DE_T}{Dt} = \frac{\partial}{\partial x_i}(\sigma_{ij} u_j). \tag{6.3}$$

In the above, σ_{ij} is the stress tensor, regarded as the sum of the hydrostatic stress, $-\delta_{ij} P$, and a stress deviator tensor, S_{ij}, so that

$$\sigma_{ij} = S_{ij} - \delta_{ij} P. \tag{6.4}$$

E_T represents the total energy (kinetic plus internal) per unit mass. Tensor notation is implied in the above. The indices *ij* have a range of 1–3 and repeated indices denote summation. See Fung [1965] or Drucker [1967] if you need a refresher.

The quantity *D/Dt* in the above expressions is referred to in the literature as the *convective derivative* or the *material derivative*. It is given as

$$\frac{D}{Dt} = \frac{\partial}{\partial t} + u_i \frac{\partial}{\partial x_i}. \tag{6.5}$$

Excellent descriptions of the physical meaning of this quantity as opposed to the Lagrangian derivative $\frac{d}{dt}$ are given by McDonald [1996] and Anderson [1995]. Referring to the above equation as the *substantial derivative*, Anderson [1995] states that "…*D/Dt* is the substantial derivative, which is physically that time rate of change following a moving fluid element; $\frac{\partial}{\partial t}$ is called the *local derivative*, which is physically the time rate of change at a fixed point; $u_i \frac{\partial}{\partial x_i}$ is called the *convective derivative*, which is physically the time rate of change due to the movement of the fluid element from one location to another in the flow field where the flow properties are spatially different. The substantial derivative applied to any flow field variable…".

Using the definition of the convected derivative and some elementary manipulations, the conservation equations may be re-written in a more convenient form for computations:

$$\frac{\partial \rho}{\partial t} = -\frac{\partial (\rho u_i)}{\partial x_i}, \tag{6.6}$$

$$\frac{\partial}{\partial t}(\rho u_j) = \frac{\partial}{\partial x_i} \sigma_{ij} - \frac{\partial}{\partial x_i}(\rho u_i u_j), \tag{6.7}$$

$$\frac{\partial}{\partial t}(\rho E_T) = \frac{\partial}{\partial x_i}(\sigma_{ij} u_j) - \frac{\partial}{\partial x_i}(\rho u_i E_T). \tag{6.8}$$

According to Hageman & Walsh [1971], "…it is desirable to replace these differential equations by the analogous integral equations, obtained by integrating over the cell volume, V, and then converting the volume integral of divergences to surface integrals over the cell surfaces. The above equations then become

$$\frac{\partial}{\partial t} \int_V \rho \, dV = -\int_S \rho u_i n_i \, dS, \tag{6.9}$$

$$\frac{\partial}{\partial t} \int_V \rho u_j \, dV = \int_S \sigma_{ij} n_i \, dS - \int_S \rho u_i u_j n_i \, dS, \tag{6.10}$$

$$\frac{\partial}{\partial t} \int_V \rho E_T \, dV = \int_S \sigma_{ij} u_j n_i \, dS - \int_S \rho u_i E_T n_i \, dS. \tag{6.11}$$

It is convenient to express the integral conservation relations above as finite-difference equations over the time step Δt, and also to decompose the total stress σ_{ij} into its deviator and hydrostatic components. This gives, for the increments of total mass (m), momenta (mu_j) and energy (mE_T) within the cell

$$\Delta m = -\Delta t \int_S \rho u_i n_i \, dS, \tag{6.12}$$

$$\Delta(mu_j) = -\Delta t \int_S P n_j \, dS + \Delta t \int_S s_{ij} n_i \, dS - \Delta t \int_S (\rho u_i u_j) n_i \, dS, \tag{6.13}$$

$$\Delta(mE_T) = -\Delta t \int_S P u_i n_i \, dS + \Delta t \int_S s_{ij} u_j n_i \, dS - \Delta t \int_S (\rho u_i E_T) n_i \, dS. \tag{6.14}$$

Here, the terms on the right are divided into increments due to pressure forces on the cell surfaces (first column), those due to the stress deviator forces on the cell surface (second column) and the increments (third column) due to the transports of mass, momentum or energy through the surface of the cell. The three types of contributions are accounted for in the computation in distinct phases. Specifically, during each time step, all cells are updated for:

- Pressure effects in Phase 1.
- Effects of the stress deviators in Phase 3.
- Transport effects in Phase 2.

The calculations are performed in the order listed, the naming of the phases having to do with the history of their development."

Euler codes developed after HELP tend to combine pressure effects with the stress deviator calculations. Thus only two phases are used in the calculation. The first pass is referred to in the literature as a Lagrange phase, the second as a rezone to the original undeformed mesh state during which material is transported from cell to cell. Perhaps this is more easily envisioned if we look at the above equations in the following way:

Euler = Lagrange + Stuff

Remembering that in a Lagrange calculation material does not pass from cell/element (there is no transport) then, in the above, the combined Phase 1 + Phase 3 terms are equivalent to a Lagrange calculation. The remaining "Stuff" then consists of material that moves in the spatial domain at any given time step. The Lagrange stage is pretty well straightforward. But, as in Lagrange codes the contact algorithm consumed most of the computing time and so made or broke the code, here the material transport methodology dominates the calculation and determines the fate of the code.

A few more comments before moving onto interface definitions and material transport. Most production Lagrange codes employ finite elements. Most Euler codes use finite differences. This is not a hard and fixed rule but a matter of historical development. Finite elements were just coming into existence when these codes were being developed. Finite-difference techniques were firmly rooted in the mathematical literature for over 100 years and many of the properties of the various difference schemes were well understood. Whereas connectivity of element is explicitly stored in finite-element codes (making possible such developments as "eroding" slide lines and the various alternative descriptions of the method of dynamically re-defining the contact surface), Euler codes operate with a fixed grid in which connectivity of cells is implicit. Thus, in developing the procedures, one focuses on an arbitrary cell "k" and its neighbors referred to as the cell to the left of k (kl), to the right of k (kr), above and below k (ka, kb) as well as above and below to the left and right of k (kal, kar, kbl, kbr). Cell numbering refers to the i,j,k position of a cell where the numbering is along the axes. The grid layout and a typical cell used in the HELP code is shown in Fig. 6.3. All subsequent Euler codes use a very similar system.

It is also important to remember that some finite-difference schemes have damping implicitly incorporated in them. The user of the code has no control over this but must be aware of its existence. Damping is also present in the various operators used for implicit calculations. More on this may be found in the classic references such as Morse & Feshbach [1953], Richtmyer & Morton [1967] and more modern works such as Belytschko & Hugers [1983] and Belytschko, Liu & Moran [2000].

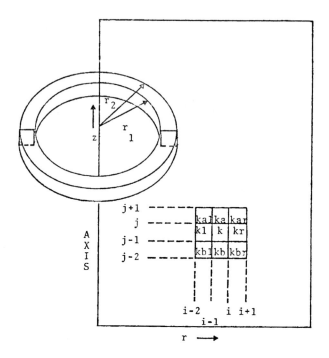

Figure 6.3 Grid layout and typical cell for the HELP Eulerian code [Hageman & Walsh, 1971].

a) *Material interfaces and transport.*

Consider Fig. 6.4(a). At the risk of boring you with repetition, remember that Lagrange codes have the mesh tied to the material. As distortions progress, problems occur with element crush-up which can render calculations inaccurate by exceeding the parameters of the material model, especially the equation of state, or reduce the time step to such a small level that further calculations become impractical. While this is going on, though, we can easily locate the free surfaces and material interfaces of the problem to great accuracy. By contrast, Euler calculations operate with a fixed mesh that has material flowing through it. As material mixing occurs, interfaces and boundaries can be determined at best to within a cell size, possibly two. In fact, in Fig. 6.4(b), I have cheated by showing you where the interface is located. A pure Euler code would simply indicate regions where cells contain more than one material. To gain a bit more accuracy we need to do something to identify boundaries. Among the earliest approaches were the use of a diffusion limiter in the HULL code [Matuska & Osborn, 1982] and the use of massless Lagrange tracer particles to define the boundaries of the various bodies involved in a calculation, Hageman & Walsh [1971].

Alternatives to Purely Lagrangian Computations   211

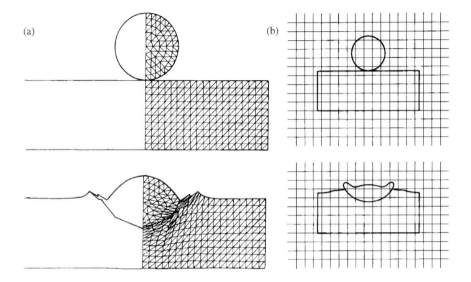

Figure 6.4 (a) Lagrangian grid; (b) Eulerian grid.

The HULL manual [Matuska & Osborn, 1982] refers to its diffusion limiter as an *ad hoc* algorithm found to be "modestly successful." At any given time step, once the Lagrange phase is over, material is moved in one coordinate direction at a time. At first, relative cell-centered velocities are computed using the donor, receiver and upstream cells. Then transport fractions are computed for each material and these are applied to mass, volume and internal energy transport from donor to receiver cell. As an example, assume a quantity of mass is to be transported to a cell to the right of the donor cell. For each material n in the donor cell, the amount of mass transport is then

$$\Delta M_n^R = (\Delta_{nR} M_n)_{\text{donor}}. \tag{6.15}$$

The total mass then fluxing to the right will be

$$\Delta M^R = \sum_n \Delta M_n^R. \tag{6.16}$$

A similar procedure is used for volume and internal energy. Analogous procedures are used for the remaining coordinate directions. The accuracy of interface and boundary locations in this procedure remains one mixed cell dimension.

The HELP code [Hageman & Walsh, 1971] used massless Lagrangian tracer particles to determine boundaries and interfaces. Then, fractional cell boundary areas are computed for each material within a mixed cell. These areas are then used in the mass transport terms. The derivation of cell boundary areas is kept fairly general, subject to the assumption that a material boundary never crosses itself. To prevent this from happening, provision was made for users to add additional tracers when the spacing became sparse. This proved to be the Achilles Heel of the code. Unless a user was very familiar with not only the code, but the physics of large deformation problems, the crossing of material boundaries became inevitable and required considerable user intervention to restore them. In turn, this excessive user influence could easily result in a solution, which mimicked the result assumed by the user rather than the physics of the situation. Nevertheless the code contained some interesting concepts for its time. Matuska has observed that codes develop by evolution, not revolution. Hence, many of the concepts in early codes are adapted and influence further code development. Thus, a quick peek at the use of tracers to define interfaces is warranted.

When an Eulerian cell (Fig. 6.5) is cut by the material N interface we need to known the fraction of the top and right side of the cell should be associated with the material N. These fractional areas will be used to compute the mass flux of material N across each side of the mixed cell. It is necessary to compute and store only the fractional areas for the right and top because we are computing (in two dimensions) on a fixed rectangular grid so that the cells below and on the left provide the information for the other two sides. The computation of fractional areas is straightforward. However, a number of decisions have to be made first:

- How many material intersections cross the mixed cell?
- In what order are materials to be transported?
- What is their contribution in terms of fractional area?

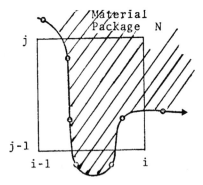

Figure 6.5 Material intersecting an Euler cell [Hageman & Walsh, 1971].

- Will a void cell be contributing material, or is transport to a void cell anticipated?
- Does the cell contain a geometric boundary?
- Will the mixed cell become a pure cell after transport?

These and many other considerations are accounted for as various cases in the program. Thus, as much time, or more, is spent with the transport algorithms in Euler codes as with contact algorithms in Lagrange codes. All interaction possibilities must be accounted for in advance lest pathological problems arise in practice. The ultimate goal is the determination of the mass flux. In HELP, the mass flux for pure cells is given by

$$\Delta M = \rho_T V_T A \Delta t. \tag{6.17}$$

In the above, ρ_T is the density of the donor cell; V_T, the average cell-centered velocity; and A, the area of the cell boundary for which mass flux is computed.

For cells crossed by an interface

$$\Delta M_n = \rho_{Tn} V_T FA_n \Delta t / S. \tag{6.18}$$

Here, the density is usually that of the donor cell material to be transported. Special treatment is required if the cells contain void interfaces. The term FA represents the fractional cell boundary area for material n. Division of the time step by S, the number of subcycles of the transport, is sometimes needed for smoothness or stability.

Details are given in the code manuals cited. The main point to be made is this. Material transport is a non-trivial matter, which is in no way automatic in these techniques. It also reflects the thinking (read bias) of code developers at the time the code was written. It is entirely possible that situation unaccounted for at the time of development may arise during the lifetime of the code. Thus, if it were not prepared in-house, a close relationship of needs must be maintained between code developers and users. This lends some weight to the assertions by Dr. Charles L. Mader [1993] at his short course on numerical modeling of detonation, to wit:

(1) The best code is the one you wrote yourself.

(2) The next best is one acquired from a reputable developer, but then a very close association is necessary between user and developer.

(3) When a code developers dies, his code should be buried with him.

The latter assertion seems especially hard. However, in the light of experience, there is much to be said about re-thinking numerical procedures every 10–15 years.

Considerable improvements in the use of Lagrangian tracers to define interfaces were made in the SMITE code [Burstein, Schecter & Turkel, 1975a,b; Burstein & Schechter, 1976a,b]. In SMITE, a second-order finite-difference code, each material has its own independent grid. Thus, the mesh spacing and the number of mesh grids in one material are not affected by any other material. The extent of each domain is determined by particles that define the domain boundary. These points were moved in Lagrangian sense by integrating the ordinary differential equations relating their positions or velocities. Additional particles were added as needed to ensure uniform spacing (and correct boundary determination) throughout the calculation. The interaction of materials was through the boundaries. The boundaries were subjected to free surface and interface conditions and provided the only communication between the various material domains in a calculation.

The next step in complexity, if not necessarily accuracy, involves the use of interface reconstruction algorithms [Scheffler & Zukas, 2000]. One of the most successful (because of its simplicity) material-interface reconstruction algorithms was the first-order accurate simple line interface calculation (SLIC) algorithm of Noh & Woodward [1976]. The algorithm treats each coordinate direction independently and represents material interfaces as straight lines either perpendicular or parallel to a coordinate direction based on the volume fractions of neighboring cells. The order in which each coordinate direction is treated is reversed each computational cycle. The SLIC algorithm did well when material flow was primarily in a coordinate direction but tended to reorient the material interfaces if material flow was not parallel to a coordinate direction. For example, SLIC would tend to reorient the material interfaces of a circle moving the mesh at 45 degrees such that it would take on a diamond shape. An example of the working of the algorithm is provided in the "balls-and-jacks" problem shown in Fig. 6.6 (a) and (b) in which two balls (resembling a "Figure 8") and two jacks are moved through the mesh at 45 degrees with x and y-velocity components of 100 km/s. the input decks and the balls-and-jacks problem used to demonstrate various interface reconstruction algorithms were generously provided by Drs. Eugene S. Hertel, Jr. and Raymond L. Bell of Sandia National Laboratories.

One such scheme is Youngs' interface reconstruction algorithm [Youngs, 1987; Parker & Youngs, 1992; Thomas & Jones, 1992]. The Youngs' algorithm also uses lines (or planes in three dimensions) to represent the material interfaces; however, the lines can be at any angle. A problem with the Youngs' algorithm is that the code user is required to input the order in which materials will advect. The wrong choice of material advection order can lead to strange results, as shown in the balls-and-jacks problem in Fig. 6.7 (a) and (b). In very complex problems involving many materials, considerable skill is needed

Alternatives to Purely Lagrangian Computations 215

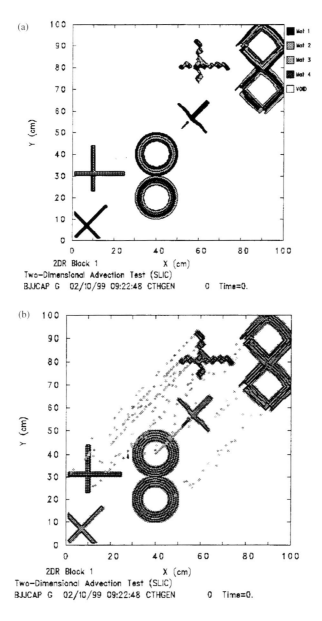

Figure 6.6 (a) The balls-and-jacks problem using the SLIC algorithm. The material interface plot shows the initial and final configurations of the balls and jacks [Scheffler & Zukas, 2000]. (b) Material volume fraction plot shows that small fractions of material are left behind [Scheffler & Zukas, 2000].

216 *Introduction to Hydrocodes*

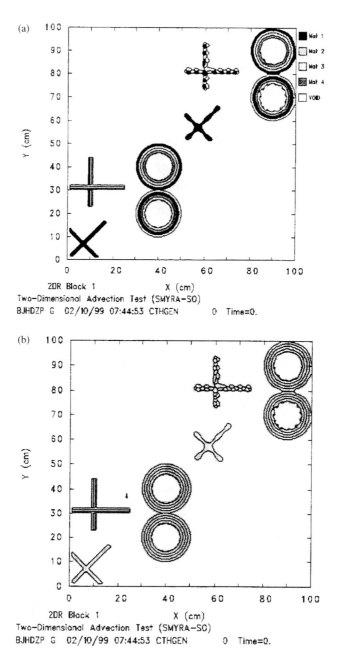

Figure 6.7 (a) The balls-and-jacks problem using Youngs' interface reconstruction algorithm with improperly chosen material advection orders. The material interface plot shows the algorithm does a better job keeping track of material interfaces than the SLIC algorithm [Scheffler & Zukas, 2000]. (b) Material volume fraction plot shows that material incorrectly advected is being left behind [Scheffler & Zukas, 2000].

in choosing the material advection order. Additionally, the advection order of materials in a complex problem is rarely uniform over the entire computational domain. The problems with the choice of material advection order were largely solved with Sandia's Modified Youngs' Reconstruction Algorithm (SMYRA) [Bell & Hertel, 1992] used in the CTH code [McGlaun & Thompson, 1990], which is essentially Youngs' algorithm, but material volume fractions of neighboring cells are used to determine the advection order in mixed cells. An example of the balls-and-jacks problem using SMYRA is shown in Fig. 6.8 (a) and (b). A comprehensive review of material reconstruction algorithms is in Benson [1992, 2002]. The interface in this case will simply disappear, and the material behaves as though it were bonded together. This occurs because material interfaces can only be determined between different materials or materials and void, not like materials (meaning materials that are assigned the same material number). An example is shown in Fig. 6.9 for a deep penetration problem involving a tungsten alloy projectile penetrating a steel target. In this case, the erosion products of the long-rod penetrator came in contact with the penetrator and the interface between the penetrator and erosion products can no longer be distinguished. The end result was that the simulation underpredicted final penetration due to the penetrator being decelerated from the contact. Velocities in most Eulerian hydrocodes are either face-centered or node-centered, while other cell quantities are cell centered. Therefore, material in a mixed cell all have the same velocity field. Thus, a cell moves as a distorting block implying a no-slip condition. Therefore, Eulerian codes do not handle very well problem where sliding between materials occurs or where material separate. Unfortunately, sliding occurs in most problems. Attempts have been made to correct this shortcoming. For example, Silling [1994] developed a boundary layer algorithm for sliding interfaces (BLINT) for 2D problems. The algorithm creates a "slip layer" in a user-defined "soft" material. The strength of the cells in the slip layer is set to zero, allowing sliding to occur outside the mixed cells. The model has been used with some success in modeling rigid-body penetration [Kmetyk & Yarrington, 1994; Scheffler, 1996; Scheffler & Magness, 1998] and has recently been extended to three dimensions. An example problem comparing simulations with and without the BLINT model is shown in Fig. 6.10 (a) and (b). Walker & Anderson [1994] attempted to overcome the problem by developing a model for multi-material velocities in mixed cells. The model defined a cell-centered velocity and allowed each material to have its own velocity, which was advected as a state variable. The authors attempted to model a rigid-body penetrator problem using the model with only limited success.

Because the cell stress increments are calculated from velocity gradients using the face-centered or node-centered velocities, a single yield strength must be calculated from mixed cells. Some Eulerian hydrocodes allow the user to select the formulation for determining the yield strength of a mixed cell. As an example, the CTH hydrocode

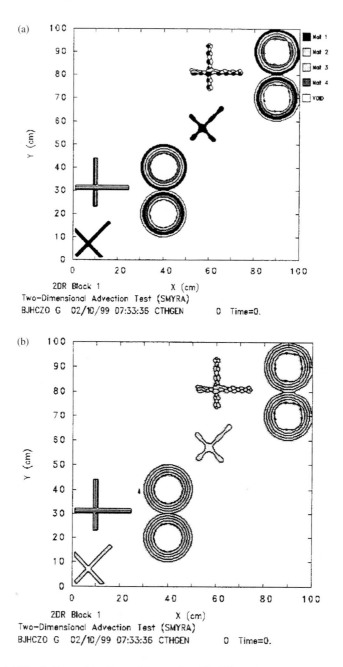

Figure 6.8 (a) The balls-and-jacks problem using SMYRA. Material interface plot shows results similar to Youngs' algorithm [Scheffler & Zukas, 2000]. (b) Material volume fraction plot shows no material left behind [Scheffler & Zukas, 2000].

Alternatives to Purely Lagrangian Computations 219

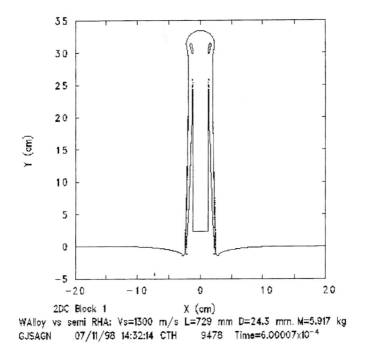

Figure 6.9 Example of a deep penetration simulation in which erosion products of the rod come into contact with the penetrator. An interface is indistinguishable at the point of contact between rod and target. The contact with erosion products increases deceleration of the rod so that penetration depth is underpredicted [Scheffler & Zukas, 2000].

gives the user three choices on how the yield strength of a mixed cell is determined. They include

- setting the strength of a mixed cell to zero;

- multiplying the yield strength of each material by its volume fraction and adding them;

- multiplying the yield strength of each material by its volume fraction, summing them and dividing by the total volume fraction of all materials within the cell.

All these methods for determining the yield strength in a mixed cell can lead to calculating unrealistic results when the materials within a mixed cell have vastly different strengths. Figure 6.11 compares simulations using the previously mentioned strength treatments for mixed cells. For each strength formulation, the penetrator did not perforate

220 *Introduction to Hydrocodes*

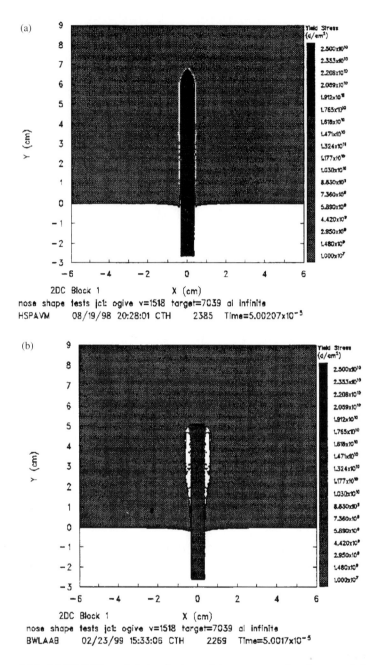

Figure 6.10 (a) BLINT model simulation of deep penetration by non-deforming penetrator [Scheffler & Zukas, 2000]. (b) Simulation without BLINT incorrectly shows rod deformation and erosion [Scheffler & Zukas, 2000].

Alternatives to Purely Lagrangian Computations 221

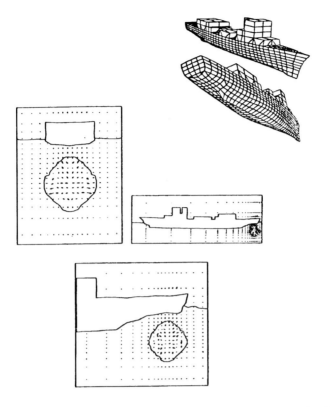

Figure 6.11 Detonation of a warhead below the first compartment of a frigate. The illustration depicts velocity contours and material interfaces [Courtesy: IABG].

the target whereas experimentally it did. The reason is partly due to the fracture model chosen. If a solid were in a mixed cell with air, the solid can undergo unrealistic deformation. The strength treatment used in mixed cells is one reason most Eulerian codes cannot model rigid-body penetration.

Mixed cells can also cause thermodynamic problems. This occurs because most Eulerian codes typically determine the thermodynamic state of the mixed cell on a sub-cell level. The thermodynamic state of neighboring cells is not used to determine the state of materials within a mixed cell. *Assumptions* must therefore be made to give a closed set of equations to determine the state within a mixed cell. The assumptions most often used include:

- All materials are at the same pressure and temperature.

- All materials are at the same pressure but the temperature of each material may vary.

- All materials may have their own pressure and temperature.

The first two assumptions are obviously incorrect. While pressures may eventually equilibrate, this usually happens over time, not instantaneously. Additionally, assuming all materials have the same pressure can cause problems to arise when fracture is to be modeled using a tension criterion in a cell containing a solid and a fluid (see Zukas [1987] for a review of failure criteria in hydrocodes). The solid will never fail because fluids do not support tension. Assuming thermal equilibrium implies an infinite thermal conductivity.

The assumption that materials have their own temperature and pressure on the surface is the most correct. However, additional approximations have to be made. These include:

- Assuming that the volumetric strain is constant among all materials in a mixed cell. This means that, in a mixed cell containing a solid and a gas, the solid would have to accept the same amount of volumetric strain as the gas in spite of the fact that the gas is much more compressible. The approximation correctly allocates *PdV* work to materials only when compression occurs parallel to a material interface. The assumption is used in the MESA, CTH and KRAKEN codes [DeBar, 1974].

- Assuming that all materials experience equal pressure change in mixed cells with multiple temperatures and pressures. The approximation correctly allocates *PdV* work among materials only when a cell is compressed perpendicular to the material interfaces and represents and opposite limiting case to constant volumetric strain approximation. This approximation has been added to the CTH code as a user-selectable option [Fransworth, 1995].

We have touched on some of the factors in Eulerian codes which can lead to disagreement between computations and experiment. It is important to remember that many *ad hoc* assumptions are introduced into material interface treatments which, when incorrectly applied, can lead to considerable error in computations. In addition, many "user options" are presented in hydrocodes, in general, which have been tested only on a very limited basis by code developers. Discovering their limitations is left to users. Code users must be aware of the problems that can arise in the treatment of material interfaces when they occur so as not to attribute them as reality or new physics. Once more it is worth repeating the old adage that in no way whatsoever can hydrocodes be considered to be "black boxes," to be picked up and readily used by junior engineers. Some six months to two years are needed to become thoroughly familiar with a code and its eccentricities. This includes expenditure of time and money to run numerous test cases to ascertain the code's response to various user options and parameter variations. In the meantime, a close relationship should be maintained with code developers and advantages taken of any and all training available on code operations and the physics and mechanics associated with fast, transient loading.

6.3   Arbitrary Lagrange–Euler methods.

Many problems can be solved using Lagrangian codes. However, as distortions increase two main problems arise, as were pointed out earlier. Severely distorted grids can lead to inaccuracies in the solution. Further, since calculations for large distortion problems use almost exclusively explicit time integration techniques, severely distorted elements reduce the time step to the point where economical calculations are impossible. Euler codes overcome both these problems. However, as they involve more calculations per time step and larger grids than Lagrangian calculations since the user must anticipate where the material will flow, these also can be expensive. Hence, there is a continuing search for methods, which will take advantage of both Euler and Lagrange techniques. This is especially true for problems where the large distortions occur in a small subdomains of the total mesh.

One such methodology is the ALE method. An excellent description of the method for finite elements is given in Chapter 7 of Belytschko *et al.* [2000]. Thus, we limit ourselves here to outlining the procedure and its advantage. It is a fertile area for research in aerodynamics and fluid–structure interaction problems.

The aim of the ALE procedure is to capture the advantages of both the Eulerian and Lagrangian methods while minimizing the disadvantages. Citing Belytschko *et al.* [2000], "...As the name suggests, ALE descriptions are arbitrary combinations of the Lagrangian and Eulerian descriptions. The word arbitrary here refers to the fact that the combinations are specified by the user through the selection of mesh motion. Of course, a judicious choice of the mesh motion is required if severe mesh distortions are to be eliminated, and this often imposes a substantial burden on the user."

What an elegant way of restating the old adage that there is no free lunch! An ALE code can alleviate some of the burden of large distortions but the price for this is a sophisticated user who is familiar with the problem he or she is trying to solve, the numerical approximations involved and considerable experience with the particular code to understand its eccentricities. There are no handbooks on mesh velocities for the various problems listed in Chapter 1. Quite a bit of trial and error is involved to develop the necessary background to establish such parameters with some degree of accuracy. More on this in Chapter 8.

While the ALE methods are quite general and can address a wide variety of problems, especially in the structural dynamics regime, an interesting interpretation for those dealing with problems involving penetration, perforation and explosive loading, is to view ALE methods as algorithms which perform automatic rezoning. Manual rezoning is performed in Lagrange codes by: (a) stopping the calculation when the unacceptable level of distortions is reached; (b) smoothing the mesh; and (c) remapping the solution from the distorted mesh to the smooth mesh. According to Stillman, Lum, Hallquist & Benson [1993], "...An ALE formulation consists of a Lagrangian time step followed by a remap or advection step. The advection step performs an incremental rezone, where incremental refers to the fact that positions of the nodes are moved only a small fraction of the characteristic lengths of the surrounding elements. Unlike a manual rezone the topology of the mesh is fixed in an ALE calculation. An ALE calculation can be interrupted like an ordinary Lagrangian calculation and a manual rezone can be performed if an entirely new mesh is necessary to continue the calculation."

Stillman *et al.*, 1993 list some additional properties of ALE calculations:

- Many ALE algorithms are second-order accurate, as compared to manual rezones which are of first-order accuracy. Thus, in theory, ALE algorithms are superior in terms of accuracy.

- In theory, an ALE formulation contains the Euler formulation as a subset. While Euler codes are not restricted to a single material, many ALE formulations are simplified to permit only one material in each element. The primary advantage of this is the reduced cost per time step. When elements with more than one material are permitted, not only can the number and type of materials in each element change every computational cycle, but additional data are necessary to specify which materials are in each element and the data must be updated in the remap algorithm.

- The range of problems which can be solved with an ALE formulation is directly related to the sophistication of the algorithms for smoothing the mesh. Stillman *et al.* list a number of different approaches to mesh smoothing. They point out that many early ALE algorithms failed because they were not successful in mesh smoothing. Most of the problems were associated with nodes on material boundaries. If the material boundaries are purely Lagrangian, no smooth mesh may be possible. The algorithms for maintaining a smooth boundary mesh are therefore as important to the robustness of the calculation as the algorithms for the mesh interior.

- Second-order transport accuracy is important in a variety of explosion and impact cases where the shock waves are weak. Errors in the transport calculations generally smooth out the solution and reduce the peak values of pressure waves. In explosive gases, the shocks are strong shocks and they are less sensitive to the advection algorithms than weak shocks. Since water behaves elastically in pressure regimes associated with explosions, the shock waves do not steepen as they propagate through the water and any diffusion of the stress wave due to the advection algorithm is permanent.

An ALE time step consists of the following:

(1) Perform a Lagrange time step.

(2) Perform an advection step.

   (a) Decide which nodes to move.

   (b) Move the boundary nodes.

   (c) Move the interior nodes.

   (d) Transport the element-centered variables.

   (e) Transport and update the momentum.

It is instructive to compare this to an Euler calculation:

(1) Perform a Lagrange Time step.

(2) Perform an advection step.

   (a) Restore nodes to original coordinates.

   (b) Transport the element-centered variables.

   (c) Transport and update the momentum.

The cost of an advection step per element is usually much greater than the Lagrangian step. Most of the time in the advection step is spent calculating the material transported between adjacent elements. Only a small part of it is spent calculating how and where the mesh should be adjusted. Stillman *et al.* [1993] suggest that "…Perhaps the simplest strategy for minimizing the cost of ALE calculations is to perform them every few time steps. The cost of an advection step is typically two to five times the cost of the

Lagrangian time step. By performing the advection step only every ten steps, the cost of an ALE calculation can often be reduced by a factor of three without adversely affecting the time step size. In multimaterial problems, it is frequently useful to perform the ALE on a material subset where most of the large distortions are likely to occur."

Stillman et al. [1993] present the ALE capability of LS-DYNA through several examples. One of these is a bar impacting a rigid surface at a velocity of 0.0227 cm/μs. The analysis was performed in a purely Lagrange mode and with an ALE step performed every 1, 4 and 10 Lagrangian steps. The pure Lagrange calculation took 5609 steps and 1132 CPU seconds while the ALE calculation performed at every tenth Lagrange step required only 1843 steps and 456 cpu seconds. Results of the two calculations were quite close.

6.4   Coupled Euler–Lagrange calculations.

Another approach to cutting computational costs and perhaps improving accuracy is to have a grid consist of both Euler and Lagrange regions and transfer information from one to the other via boundary conditions. There is no grid motion superimposed on the calculation as in the ALE approach. Rather, one or more Euler–Lagrange boundaries are specified and information is transferred through them.

By way of illustration, consider Fig. 6.11. This depicts a DYSMAS code (see Zukas [1990] for a summary of code capabilities circa 1990) calculation of the response of a ship to an underwater blast. The explosive was modeled as a 1D energy source. The governing equations for the energetic material follow the descriptions given by Mader [1979]. This 1D energy source was embedded in a 2D Eulerian grid modeling the water. Pressures from the explosive were transferred to the Euler grid through the explosive–water boundary. The pressures were propagated until eventually reaching the 3D Lagrangian grid modeling the ship. At each time step the Eulerian grid (the water) delivers a load to the Lagrangian interface elements (the ship structure). Ship deformations are computed, including changes to the water–ship interface. This ever-changing interface then acts as a wall boundary to the water, allowing fluid motion only in a tangential direction to the interface.

This type of approach, where one body (here, the ship structure) is much stiffer than the other (the water) requires a more elaborate time-step control than has a simple explicit scheme. Subcycling is usually included to ensure computational efficiency. The coupling procedure in DYSMAS is useful for *fluid–structure interaction* problems such as vessels subjected to underwater detonation, including submerged structures; *solid–structure interaction* problems such as building loaded by buried charges; and *impact and penetration* problems involving soil or armor penetration.

Other examples of coupled calculations are given by Chisum & Shin [1995] and Birnbaum, Francis & Gerber [1998].

Each code has a somewhat different approach to coupling Lagrange and Euler grids. How it is done is determined by the problem that one is attempting to solve. The following description of stress coupling in the HULL code is provided through the courtesy of John Osborn of Orlando Technology Inc.

The OTI * HULL wave propagation computer program is a proprietary version of the HULL code. It is proprietary to General Dynamics (Ordnance and Tactical Systems) located in Niceville, FL. Authors of the code include several scientists now at GD-OTS: Daniel Matuska, John Osburn, Edwin Piburn III and Michael Gunger. The code can be run in one of several modes—fully Eulerian, fully Lagrangian, interactively linked Euler and Lagrange and stress or velocity linked. In the interactively linked mode Lagrangian regions exist inside an overall Eulerian solution space and both Euler and Lagrange modules run alternately with boundary conditions from the other module. In a stress or velocity link a fully Eulerian problem is run with time history stations placed on the periphery of the region to eventually be used as input to the lagrangian module. In the Euler run this region is a normal Eulerian region (i.e., composed of deformable materials) for a velocity link and a non-deformable body for a stress link.

The non-deformable (rigid) body translates and rotates based on the forces applied to it in the Euler run and according to input values of mass, initial velocities, centers-of-gravity and moments-of-inertia. Zones inside the rigid body are identified in the code and no calculations are performed for them. However, computations are made to define the rigid-body motion each cycle. The rigid body itself does not move but the rest of the grid is translated and rotated in response to the calculated motion.

Material which impacts the rigid body is presumed to stagnate in the direction of motion. This means that the solution is completely valid only for rigid bodies, which represent materials with large shock impedances compares to the Euler target materials. Although this introduces a slight limitation, this code feature has found great use in the simulation of steel projectiles impacting soil, concrete or rock.

An example of such an Euler calculation is seen in Figs. 6.12–6.15. These are plots at 0, 2.5, 5 and 12 ms of a rigid penetrator impacting concrete at 950 fps. Actually, as mentioned earlier, the penetrator is stationary and the target is moved against it. A plug forms in this finite target as the penetrator exists, Fig. 6.15.

This calculation is useful in itself since the code provides decelerations, velocities and penetration depths as the problem runs. A plot of velocity vs. penetration from the calculations is provided in Fig. 6.16. The slope of the curve changes radically as target material begins to move at the back surface. Deceleration vs. penetration is plotted in Fig. 6.17. The peak deceleration of about 12,000gs occurs when the penetrator's ogive nose becomes completely buried in the concrete.

228 *Introduction to Hydrocodes*

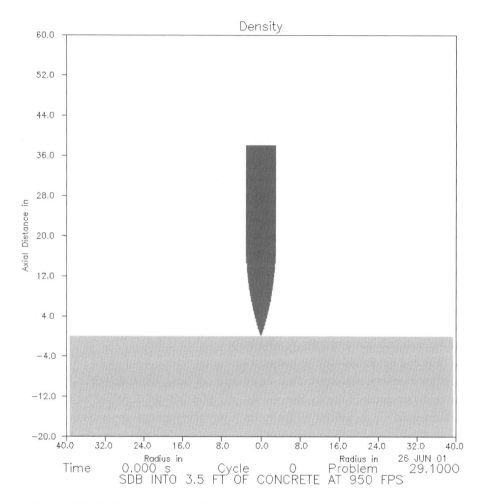

Figure 6.12 Euler calculation of concrete penetration by a rigid-body projectile [Courtesy: John Osborn, GD-OTS].

For a linked run, time history stations can be placed around the periphery of the rigid penetrator during the Eulerian run. Since the penetrator is fixed in the Eulerian grid, Eulerian time history stations (fixed also in the grid) are used to store stresses. This time history file can then be used to provide stress boundary conditions for a Lagrangian run.

The penetrator is modeled in the Lagrangian module as seen in Figs. 6.17 and 6.18. Figure 6.17 displays the Lagrangian grid and Fig. 6.18 provides material shading. The darkest area is explosive and a battery and the lightest area is a fuze model. The intermediate gray scale is the steel case. Note that this Lagrangian model is longer than the rigid body used in the Euler run. This does not present a problem as long as the

Alternatives to Purely Lagrangian Computations 229

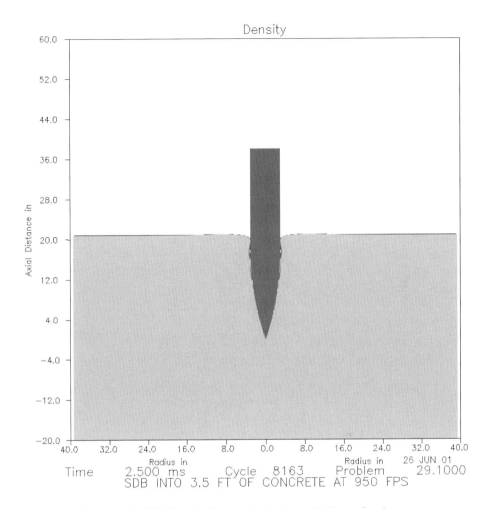

Figure 6.13 HULL code Euler calculation at 2.5 ms after impact
[Courtesy: John Osborn, GD-OTS].

masses, centers-of-gravity, etc. are the same. There only need be enough rigid body penetrator to provide the shape of the loaded surface. Since nothing but the Ogive nose is loaded in this axisymmetric calculation the nose is the only section that needed to be carried in the Euler mesh.

Element sides on the surfaces of the Lagrangian penetrator are identified by the user as linked sides. Time history data from the Euler run are then used to provided stresses vs. time for these sides as the calculation progresses.

Figure 6.19 plots the Lagrangian grid at 2.9 ms after impact. The penetrator has gone through a slightly plastic phase at the point where the tapered steel case necks down

230 *Introduction to Hydrocodes*

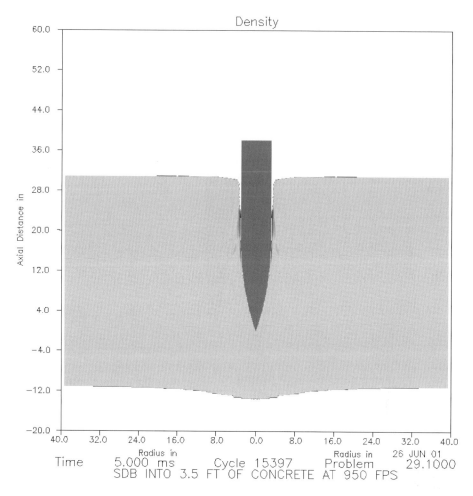

Figure 6.14 HULL results at 5 ms after impact [Courtesy: John Osborn, GD-OTS].

to the thickness of the cylindrical section. The plastic region is highlighted in Fig. 6.20 with the number of cycles a zone has been plastic (up to 99) printed over the material outline (Fig. 6.21).

Besides investigating case integrity stations can be placed in the case wall and the fuze simulant to provide detailed stress plots and decelerations of interest.

Velocity boundary conditions were used in an earlier study to devise a feasible computational methodology for performing calculations of long-rod penetrators perforating spaced target arrays [Matuska & Osborn, 1981]. Such calculations are most challenging, even today, in their requirement of computational time. As the authors

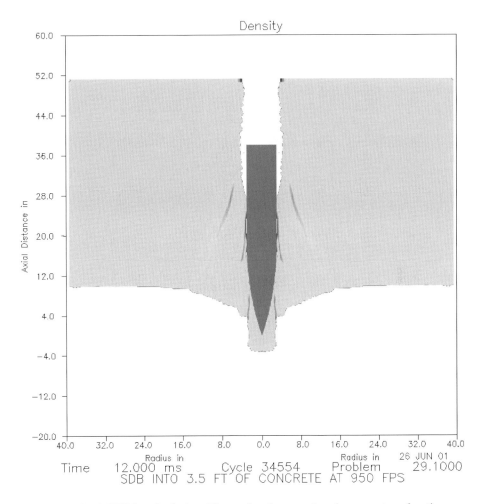

Figure 6.15 HULL calculation 12 ms after impact showing target perforation [Courtesy: John Osborn, GD-OTS].

point out, "...A three-dimensional calculation will require on the order of N times as much computer time and storage, where N is the number of discretized elements in the added dimension; thus, two-dimensional calculations, which require an hour of central processor time and 100,000 storage elements, would require several ten's of hours and several millions of words of storage in three dimensions. From a practical point of view, the ability to complete these calculations is dominated by the economics of computing, and by the amount of time required to complete a calculation and interpret the results."

The study involved experiments, numerical simulations and material characterization at high strain rates, all of which are nicely detailed by Matuska & Osborn [1981]. As a demonstration of the technique, a calculation was run with the 3D HULL code until

232 Introduction to Hydrocodes

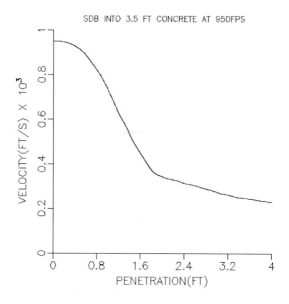

Figure 6.16 Velocity vs. penetration depth for the rigid body projectile [Courtesy: John Osborn, GD-OTS].

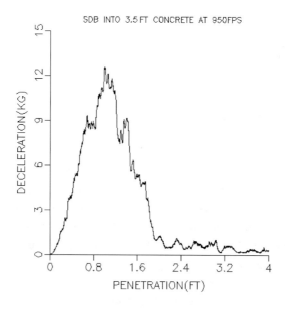

Figure 6.17 Deceleration vs. penetration depth for the rigid body projectile [Courtesy: John Osborn, GD-OTS].

Alternatives to Purely Lagrangian Computations 233

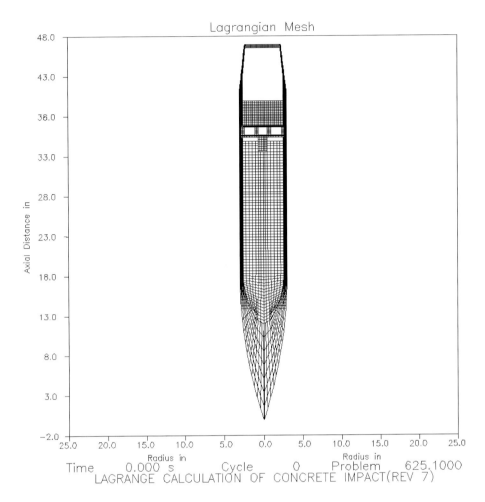

Figure 6.18 Lagrangian zoning [Courtesy: John Osborn, GD-OTS].

penetration of the first plate of a multi-plate array occurred, as shown in Fig. 6.22. Velocity–time histories were collected at various locations on the penetrator.

The calculation was then re-run on a special version of the EPIC code for which logic had been developed for linking with HULL [Osborn, 1981]. In this calculation, the target plates were not modeled. Only the calculator was included and it was driven by the velocity–time histories collected in the previous HULL calculation.

HULL data was used to drive the rod in the Lagrange calculation for 49 μs, the time required to perforate the first plate. From then on until 200 μs, the estimated time for contact with the second plate of the spaced plate array, EPIC ran without HULL input. Figure 6.23 shows a montage of velocity vector views of the rod in its plane of symmetry.

234 *Introduction to Hydrocodes*

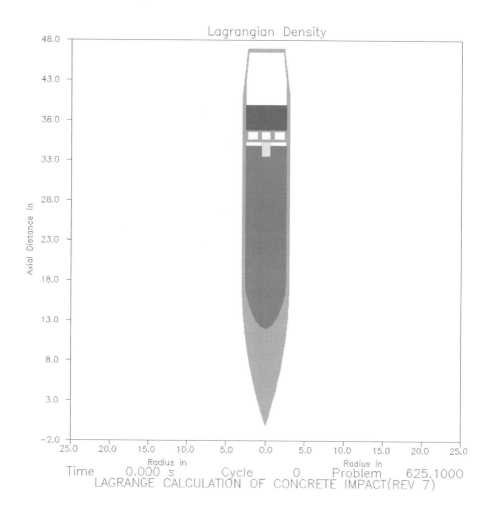

Figure 6.19 Lagrangian materials [Courtesy: John Osborn, GD-OTS].

The rod is oriented in the figure so as to properly demonstrate the rotation seen in the calculation.

Comparison with experiment for rod mass, velocity and orientation was very good, Fig. 6.24. The HULL run to perforate the first place and collect velocity–time data required 8 h of CPU time on a leading mainframe of the time. It is estimated that using HULL alone to reach the second place by 200 μs would have required 35 cpu hours. The EPIC calculation required 1 h. Thus, the linked calculation required a total of 9 cpu hours or about 1/4 of the time required for a pure Eulerian calculation with HULL. At the time the Lagrangian EPIC calculation was performed, neither rezoners nor eroding slide-line logic was available. Thus, it was unlikely that EPIC alone could have done the job owing to the severe deformation if the rod and plate elements encountered in the perforation process.

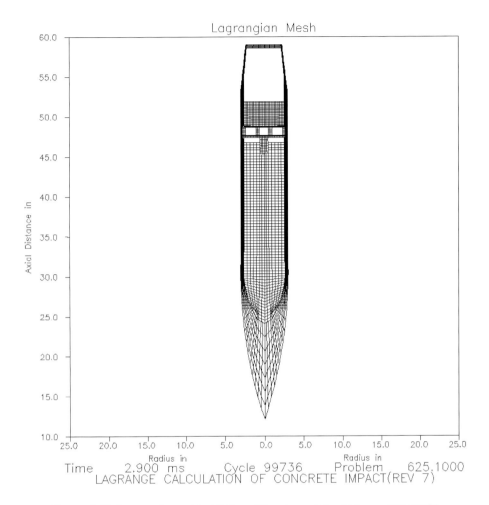

Figure 6.20 Lagrangian grid at 2.9 µs [Courtesy: John Osborn, GD-OTS].

6.5 Smoothed-Particle hydrodynamics (SPH).

All the techniques, Euler and Lagrange, discussed thus far involve the use of a computational grid. This is both good and bad. A grid provides structure and order, which can be exploited in numerical schemes. The grid is also the source of numerical pitfalls. Lagrange grids distort giving rise to errors as well as minuscule time steps. Transport through Euler grids can give rise to situations where tiny amounts of material, occupying a volume between one-thousandth to one-millionth of the cell volume, cause problems in equilibrating equation of state parameters. Hundreds to thousands of iterations can be performed to equilibrate density, pressure and internal energy in regions of the

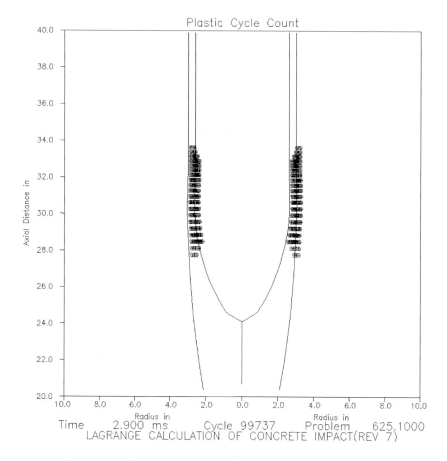

Figure 6.21 Plastic zones [Courtesy: John Osborn, GD-OTS].

calculation, which are of no great interest and have little effect on the final result. Every so often, formally or informally, scientists have observed that the particles in the flow field know where to go. The problems come about when we try to connect them with straight lines. However, in the solid mechanics arena at least, no follow-up work was done.

The SPH technique uses no underlying grid. It is a pure Largrange particle method originally developed by Lucy [1977], Gingold & Monaghan [1977, 1982], Monaghan & Gingold [1983], Monaghan [1982, 1988, 1992] and Benz [1990] for the study of stellar dynamics. A significant contribution to the SPH literature was the addition of strength effects by Libersky & Petscheck [1990]. Axisymmetric algorithms were developed by Johnson, Petersen & Stryk [1993] as well as Petscheck & Libersky [1993]. SPH nodes have been linked to finite elements by Johnson [1994] and Attaway, Heinstein & Swegle [1994].

Alternatives to Purely Lagrangian Computations 237

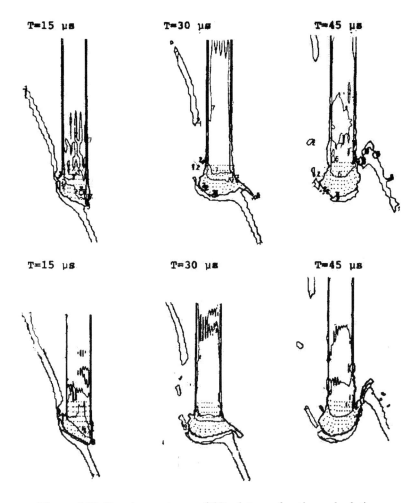

Figure 6.22 Density contours of 3D plate perforation calculations [Matuska & Osborn, 1981].

The method has two major attractions:

(1) The absence of a mesh and the calculation of interactions among particles based on their separation alone means that large deformations can be computed without difficulty.

(2) The Lagrangian formulation of SPH allows it to be straightforwardly linked to standard finite-element Lagrange formulations. This implies that both structural dynamics and wave propagation problems could in principle be computed with a single, well-constructed Lagrange code with elastic deformations and very large

238   Introduction to Hydrocodes

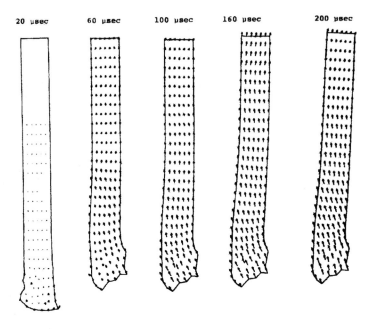

Figure 6.23 EPIC velocity vector plots to 200 μs [Matuska & Osborn, 1981].

deformations leading to material failure and separation could live side by side in the same calculation.

Before running naked into the streets and shouting "Eureka," keep in mind the oft-cited warning in this book about not expecting free lunches. The technique is very promising in theory, yes. But a hint that just a tad more research may be needed comes from Johnson, Stryk & Beissel [1996]: "…A long term objective is to allow the user to define almost any impact problem in a standard finite-element grid, and then to allow the standard finite elements to be converted to SPH nodes as the standard finite elements become distorted. Although this approach has been demonstrated, more work is required to increase the accuracy and robustness for a wider range of problems."

Current SPH formulations (e.g., see Libersky *et al.* [1993]; Randles & Libersky [1996]; Clegg *et al.* [1997]; Hayhurst, et al. [1996]; Atluri & Shen [2002]) employ an explicit time marching algorithm to advance the solution from some time t to a time $t + \delta t$. Let us use problems in solid mechanics for our example. Particle properties of density (ρ), velocity (U), energy (E), deviatoric stress (S) and position (x) are computed from the following equations:

$$\rho^{n+1} = \rho^n (1 - D\delta t^n), \tag{6.19}$$

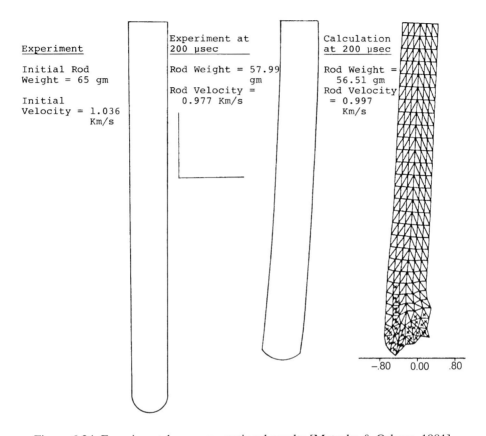

Figure 6.24 Experimental vs. computational results [Matuska & Osborn, 1981].

$$U_\alpha^{n+1/2} = U_\alpha^{n-1/2} + \frac{1}{2}(\delta t^n + \delta t^{n-1})F, \tag{6.20}$$

$$E^{n+1} = E^n + \delta t^n G, \tag{6.21}$$

$$S_{\alpha\beta}^{n+1} = S_{\alpha\beta}^n + \delta t^n H, \tag{6.22}$$

$$x_\alpha^{n+1} = x_\alpha^n + U_\alpha^{n+1/2} \delta t^n. \tag{6.23}$$

In the above, the $n+1$ superscript represents the solution at the advanced time $t + \delta t^n$. The n superscript identifies data from the previous integration cycle t. The time step identified by δt^n represents the time step that is used to advance the solution to the new time while δt^{n-1} identifies the time step used in the previous integration cycle. The time step in computed by finding the minimum $wh/(c+s)$ over all particles. Here c is the sound speed, s is the particle velocity, h is the smoothing length used in the interpolation functions and w is a stability factor between 0 and 1. Schraml & Kimsey [1998] assert that a value of 0.3 is typical. The subscripts α and β in the above equations denote spatial directions. When used individually they serve as vector notation for positions and velocities. When used together they identify stress components.

So far, so good. Here comes the fun part. The variables D (volumetric strain), F (total acceleration), G (work per unit mass) and H (stress rate) are computed from an interpolation function, resulting in "kernel estimates" of field variables at a point. To compute the new values of the field variables for a particle, that particle's previous values and the values of its neighboring particles are used with a smoothing function. A very popular one currently is the B-spline smoothing function below:

$$W = \frac{15}{7}(\frac{2}{3} - v^2 + \frac{1}{2}v^3) \quad if \quad 0 < v < 1, \qquad (6.24)$$

$$W = \frac{5}{14}(2-v)^3 \quad if \quad 1 < v < 2, \qquad (6.25)$$

$$W = 0 \quad if \quad v > 2. \qquad (6.26)$$

Here, $v = |x_i - x_j|/h$. This smoothing function generates a weighting factor for the values of neighboring particles which is based on the smoothing length, h, and the absolute distance of the neighboring particle from the particle of interest, $|x_i - x_j|$. This interpolation function is always positive, maximum at a distance of zero and reaches zero at a distance of $2h$. Thus, to compute the new interpolation functions for a particle, the only neighboring that have an influence are those within a distance of two smoothing lengths of interest. The B-spline function discussed above is plotted in Fig. 6.25. Alternatives are discussed in the cited literature.

The key phrase in the above is kernel estimates. How do we get these and why do not we need them in conventional Lagrange and Euler calculations discussed so far? Remember that grid-based methods assume a connectivity between nodes to construct spatial derivatives. SPH uses a kernel approximation which is based on randomly distributed interpolation points with no assumptions about which points are neighbors in order to calculate spatial derivatives. In this method, the continuum is represented by a set of interacting particles. This is shown in Fig. 6.26.

Each particle "I" interacts with all other particles "J" which lie within a given distance from it. This distance is called the smoothing length, h. In current SPH codes, this is specified by the user. Since SPH is still a novel technique for solid mechanics applications, the effects of choosing different values of h are not generally

Alternatives to Purely Lagrangian Computations   241

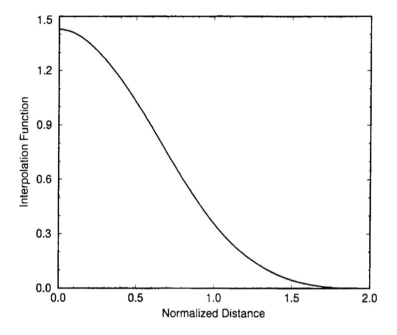

Figure 6.25 The interpolation smoothing function [Schraml & Kimsey, 1998].

known. The same holds for other free parameters in the method. Thus, different users using different values of free parameters can come up with different results for the same problem. This is a good time to recall Cauchy's comment about being able to draw a

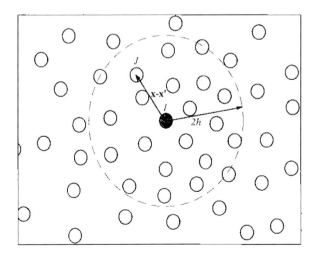

Figure 6.26 Neighboring particles of a kernel estimate [Hayhurst *et al.*, 1996].

credible version of an elephant with four free parameters and making its tail wiggle if given five!

The interaction is weighted by the function W which is called the smoothing (or kernel) function. The set of particles whose values must be used to compute the new state variables for a given particle depends on the position of the particles. Because particle positions are updated at every time step, a list of neighboring particles must be generated for each particle for every integration cycle. The smoothing function defines the set of neighboring particles that may interact with a given particle. Because the smoothing function depends on the smoothing length h, a small smoothing length will result in fewer particles in the neighbor list, and vice versa. Thus far only a small number of smoothing functions have been tried in SPH calculations.

The value of a continuous function, or its derivative, can be estimated for any particle "I" based on known values of the surrounding particles "J". In the AUTODYN code (Hayhurst et al., 1996], the kernel estimates are obtained from:

$$f(\mathbf{x}) = \int f(\mathbf{x}')W(\mathbf{x} - \mathbf{x}', h)d\mathbf{x}', \qquad (6.27)$$

$$\nabla \cdot f(\mathbf{x}) = \int \nabla \cdot f(\mathbf{x}')W(\mathbf{x} - \mathbf{x}', h)d\mathbf{x}', \qquad (6.28)$$

where f is a function of the 3D position vector $\mathbf{x}$; $d\mathbf{x}'$ is a volume.

In this procedure, the density is calculated from the continuity equation in the following way:

$$\rho \cdot = -\sum_{J=1}^{N} m^j \sum_{j=1}^{3} (v_j^I - v_j^J) \frac{\partial W}{\partial x_j^J}. \qquad (6.29)$$

Relationships for strain rate, viscosity, the constitutive relation and forces follow in a similar way. These are given by Hayhurst et al. [1996], Libersky et al. [1993] and other cited sources.

For every increment in time, each particle must compare its position to all other particles in the computation and must build a neighbor list before the state variables can be updated. For a calculation containing N particles, this implies a workload of $N * N$ position comparisons. Because this is a time-consuming process, the MAGI code [Libersky et al., 1993] generates an artificial rectilinear grid over the computational space. Particles are then assigned to cells of this artificial grid. The cells within the artificial grid have sides of length $2h$, the range of the smoothing function. For a particular time step, a particle may only interact with, at most, the other particles in its artificial cell and those in the immediate neighbor cells. This approach reduces the workload of constructing neighbor lists from $N * N$ to approximately $N \log(N)$.

On the whole, the computational costs of SPH methods are below those of conventional grid-based Euler and Lagrange techniques. However, as Hayhurst *et al.* [1993] point out, "...Whilst SPH is relatively immature, it is likely that recent and ongoing work in this active field will result in considerable improvements so the method and SPH techniques will become more widely used and established." Translation: *caveat emptor*—let the user beware—for now at least.

As an example of the application of SPH to problems of practical interest, consider the penetration of a semi-infinite target by a disk-shaped penetrator [Schraml & Kimsey, 1988]. The problem had previously been addressed with conventional Lagrange and Euler codes [Bjerke, Zukas & Kimsey, 1992]. Thus there is a basis of comparison between all three methods as well as experiments.

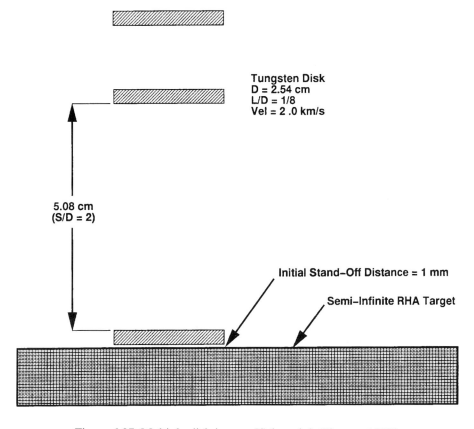

Figure 6.27 Multiple disk impact [Schraml & Kimsey, 1998].

244 *Introduction to Hydrocodes*

The SPH calculation simulated the impact of a single and spaced multiple disks of tungsten alloy striking a rolled homogeneous armor (steel) plate at a velocity of 2 km/s. Material properties for striker and target are given in Schraml & Kimsey [1998]. The initial configuration for the calculations is shown in Fig. 6.27. We will cite only the single disk impact results here.

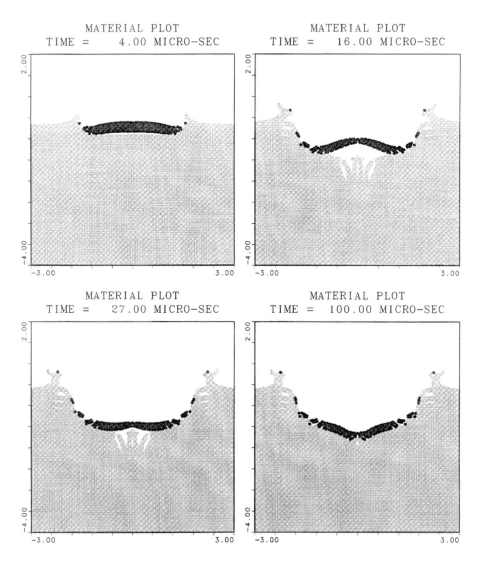

Figure 6.28 Single-disk impact at 2, 16, 27 and 100 μs ($h = 0.035$ cm) [Schraml & Kimsey, 1998].

A smoothing length of $h = 0.0635$ cm, with one particle per smoothing length, was used in the calculations. The disk thickness was 0.3175 cm. This resulted in a distribution of 5 particles across the thickness of the disk and 20 particles across the radius, for a total of 100 particles in the disk. The target was modelled with the same smoothing length and particle size as the disk and was made up therefore of 24,649 particles. The disk was initially positioned 1 mm from the impact face of the target. The calculation was run to a simulated time of 100 μs requiring 2527 time iterations on an SGI Origin 2000 System.

Figure 6.28 shows disk and target deformation at various times. Figure 6.29 shows the position of tracers on both disk and target as functions of time. The final computed penetration depth was 1.34 cm. Additional calculations employing 10, 15 and 20 particles across the disk thickness produced penetration depths of 1.43 cm. The authors conclude that the use of 5 particles to model the critical dimension, the disk thickness, is insufficient to properly capture the physics of the penetration event. The results also showed that there is no benefit to using more than 10 particles across the disk thickness of this problem. A total of four tests were performed at this velocity [Bjerke, Zukas & Kimsey, 1992]. These gave a mean penetration depth of 1.53 cm with a standard deviation of 0.26 cm. Computations using 5 particles across the disk thickness fall outside the data range while those with 10 or more fall within the data scatter with a final penetration depth that is 6.5% less than the experimental mean. The results of the SPH calculations are consistent with those obtained with conventional Euler and Lagrange methods.

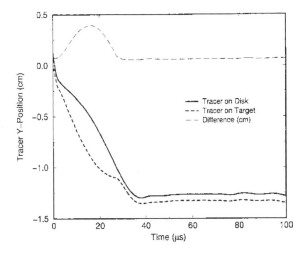

Figure 6.29 Tracer position history for single-disk impact ($h = 0.035$ cm) [Schraml & Kimsey, 1998].

REFERENCES

J.D. Anderson Jr. (1995). *Computational Fluid Dynamics: The Basics with Applications*. New York: McGraw-Hill.

S.N. Atluri and S. Shen. (2002). *The Meshless Local Petrov–Galerkin (MLPG) Method*. Encino, CA: Tech Science Press.

S.W. Attaway, M.W. Heinstein and J.W. Swegle. (1994). Coupling of smooth particle hydrodynamics with the finite element method. *Nucl. Engng. Des.*, 150, 199-205.

R.L. Bell and E.S. Hertel, Jr. (1992). An improved material interface reconstruction algorithm for Eulerian codes, Sandia National Laboratories, Report SAND92-1716.

T. Belythscko and T.J.R. Hughes (eds). (1983). *Computational Methods for Transient Analysis*, Computational Methods in Mechanics, vol. 1. Amsterdam: North-Holland.

T. Belytschko, W.K. Liu and B. Moran. (2000). *Nonlinear Finite Elements for Continua and Structures*. New York: Wiley.

D.J. Benson. (1992). Computational methods in Lagrangian and Eulerian hydrocodes. *Comput. Methods Appl. Mech. Engng*. 99, 235-394.

D.J. Benson. (2002). Volume of fluid interface reconstruction methods for multi-material problems, *Applied Mechanics Reviews*, 2(55), 151-165.

W. Benz. (1990). *Numerical Modeling of Nonlinear Stellar Pulsation: Problems and Prospects*. Boston: Kluwer Academic.

N.K. Birnbaum, N.J. Francis and B.I. Gerber. (1998). Coupled techniques for the simulation of fluid-structure and impact problems. In H.S. Levine, J.A. Zukas, D.M. Jerome and Y.S. Shin, eds, *Structures Under Extreme Loading Conditions—1998*, PVP-vol. 361. New York: ASME.

T.W. Bjerke, J.A. Zukas and K.D. Kimsey. (1992). Penetration performance of disk shaped penetrators. *Int. J. Impact Engng*, 12, 263-280.

S.Z. Burstein and S.H. Schechter. (1976a). A second order numerical code for plane flow approximation of oblique impact, USA Ballistic Research Laboratory, Report BRL-CR-294.

S.Z. Burstein and H.S. Schechter. (1976b). Fracture modeling in the SMITE code, USA Ballistic Research Laboratory, Report BRL-CR-295.

S.Z. Burstein, H.S. Schechter and E.L. Turkel. (1975a). A second order numerical model for high velocity impact phenomena, USA Ballistic Research Laboratory, Report BRL-CR-239.

S.Z. Burstein, H.S. Schechter and E.L. Turkel. (1975b). SMITE—a second order Eulerian code for hydrodynamic and elastic–plastic problems, USA Ballistic Research Laboratory, Report BRL-CR-255.

J.E. Chisum and Y.S. Shin. (1995). Coupled Lagrangian–Eulerian and multimaterial Eulerian analysis in underwater shock research. In Y.S. Shin, ed, *Structures Under Extreme Loading Conditions*, PVP-vol. 299. New York: ASME.

R.A. Clegg, J. Sheridan, C.J. Hayhurst and N.J. Francis. (1997). The application of SPH techniques in AUTODYN-2D to kinetic energy penetrator impacts on multi-layered soil and concrete targets. In *Proceedurs of the 8th International Symposium on Interaction of the Effects of Munitions with Structures*, 22–25 April 1997, VA.

R.B. DeBar. (1974). Fundamentals of the KRAKEN code, Lawrence Livermore National Laboratory, Report UCID-17366.

J.K. Dienes, M.W. Evans, L.J. Hageman, W.E. Johnson and J.M. Walsh. (1968). An Euler method for calculating strength dependent deformation, General Atomic Corp, Report GAMD-8497, Parts I, II and III and Addendum (AD 678565, 678566, 678567, 678568), February.

D.C. Drucker. (1967). *Introduction to Mechanics of Deformable Solids*. New York: McGraw-Hill.

T.E. Faber. (1995). *Fluid Dynamics for Physicists*. Cambridge: Cambridge U. Press.

A.V. Fransworth, Jr. (1995). CTH reference manual: cell thermodynamics modifications and enhancements, Sandia National Laboratories, Report SAND95-2394.

Y.C. Fung. (1965). *Foundations of Solid Mechanics*. Englewood Cliffs, NJ: Prentice-Hall.

R.A. Gingold and J.J. Monaghan. (1977). Smoothed particle hydrodynamics: theory and application to non-spherical stars. *Mon. Not. R. Astron. Soc.*, 181, 375-389.

R.A. Gingold and J.J. Monaghan. (1982). Kernel estimates as a basis for general particle methods in hydronomics. *J. Comput. Phys.*, 46, 429-453.

L.J. Hageman and R.G. Herrmann. (1977). Incorporation of the NAG-FRAG model for ductile and brittle failure into HELP, a 2D multi-material Eulerian program, Systems, Science and Software, Inc., SSS-R-77-3389, September.

L.J. Hageman and J.M. Walsh. (1971). HELP, a multi-material Eulerian program for compressible fluid and elastic–plastic flows in two space dimensions and time, Volume I, USA Ballistic Research Laboratory, Report BRL-CR-39, May.

F. Harlow and M. Evans. (1957). The particle-in-cell method for hydrodynamics calculations, Los Alamos Scientific Laboratory, Report LA-2139.

C.J. Hayhurst, R.A. Clegg and I.H. Livingstone. (1996). SPH techniques and their application in high velocity and hypervelocity normal impacts, *Proc. TTCP WTP 22nd Annual General Mtg, Hydrocode Workshop*, April 10–15, 1996, Banff, Alberta, Canada.

W. Herrmann, L.D. Bertholf and S.L. Thompson. (1974). Computational methods for stress wave propagation in nonlinear solid mechanics, Sandia National Laboratories, Report SAND-74-5397.

W.E. Johnson. (1965). OIL, A continuous, two-dimensional eulerian hydrodynamic code, General Atomic Corp, Report GAMD-5580 Revised, January.

G.R. Johnson. (1994). Linking of Lagrangian particle methods to standard finite element methods for high velocity impact computations. *Nucl. Engng Des.*, 150, 265-274.

N.L. Johnson. (1996). The legacy and future of CFD at Los Alamos, Los Alamos National Laboratory, Report LA-UR-96-1426.

W.E. Johnson and C.E. Anderson Jr. (1987). History and application of hydrocodes in hypervelocity impact. *Int. J. Impact Engng*, 5(1–4), 423-440.

G.R. Johnson, E.H. Petersen and R.A. Stryk. (1993). Incorporation of an SPH option in the EPIC code for wide range of high velocity impact computations. *Int. J. Impact Engng*, 14, 385-394.

G.R. Johnson, R.A. Stryk and S.R. Beissel. (1996). SPH for high velocity impact computations. *Comput. Methods Appl. Mech. Engng*, 139, 347-373.

L.N. Kmetyk and P. Yarrington. (1994). Analysis of steel rod penetration into aluminum and concrete targets with comparison to experimental data, Sandia National Laboratories, Report SAND94-1498.

L.D. Libersky and A.G. Petschek. (1990). Smooth particle hydrodynamics with strength of materials. In *Advances in the Free Lagrange Method*. Lecture Notes in Physics, vol. 395, pp. 248–257, NY: Springer-Verlag.

L.D. Libersky, A.G. Petschek, T.C. Carney, J.R. Hipp and F.A. Allahdadi. (1993). High strain rate Lagrangian hydrodynamics: a three-dimensional SPH code for dynamic material response. *J. Comput. Phys.*, 109, 1.

L.B. Lucy. (1977). A numerical approach to the testing of fusion process. *Astron. J.*, 83, 1013-1024.

C.L. Mader. (1979). *Numerical Modeling of Detonation*. Berkley: U. of California Press.

C.L. Mader. (1993). *Detonation Physics*, course notes, Baltimore: Computational Mechanics Associates.

G.T. Mase and G.E. Mase. (1999). *Continuum Mechanics for Engineers*, 2nd ed. Boca Raton: CRC Press.

D.A. Matuska and J.J. Osborn. (1981). HULL/EPIC3 linked Eulerian/Lagrangian calculations in three dimensions; joint report of the USA Ballistic Research Laboratory and Eglin AFB, Contract Report ARBRL-CR-00467/AFATL-TR-71-81, September 1981.

D.E. Matuska and J.J. Osborn. (1982). HULL Documentation. Volume I. Technical Discussion, Orlando Technology Inc.

P.H. McDonald. (1996). *Continuum Mechanics*. Boston: PWS Publishing Co.

J.M. McGlaun and S.L. Thompson. (1990). CTH: a three-dimensional shock wave physics code. *Int. J. Impact Engng*, 10(1–4), 351-360.

J.J. Monaghan. (1982). Why particle methods work. *SIAM J. Sci. Stat. Comput.*, 3, 422-433.

J.J. Monaghan. (1988). An introduction to SPH. *Comput. Phys. Commun.*, 48, 89-96.

J.J. Monaghan and R.A. Gingold. (1983). Shock simulation by the particle method SPH. *J. Comput. Phys.*, 52, 374-389.

J.J. Monaghan. (1992). Smoothed particle hydrodynamics, *Ann. Rev. Astronomy Astrophys.*, 30, 543-574.

P.M. Morse and H. Feshbach. (1953). *Methods of Theoretical Physics*, Parts I and II. New York: McGraw-Hill.

W.F. Noh and P. Woodward. (1976). SLIC (Simple line interface calculation). In A.I. Van de Vooren and P.J. Zandbergen, eds, *Proceedings of the 5th International Conference on Numerical Methods in Fluid Dynamics*, Lecture Notes in Physics, vol. 59. Berlin: Springer-Verlag.

J.J. Osborn. (1981). Improvements in EPIC3—link to HULL—improved equation of state—improved fracture modeling—SAIL update system—other improvements, USAF Eglin AFB, Report AFATL-TR-81-60, June 1981.

B.J. Parker and D.L. Youngs. (1992). Two and three dimensional Eulerian simulation of fluid flow with material interfaces, UK Atomic Weapons Establishment, AWE Preprint 01/92.

A.D. Petscheck and L.D. Libersky. (1993). Cylindrical smoothed particle hydrodynamics. *J. Comput. Phys.*, 109, 76-83.

P.W. Randles and L.D. Libersky. (1996). Smoothed particle hydrodynamics: some recent improvements and applications. *Comput. Methods Appl. Mech. Engng*, 139, 375-408.

R.D. Richtmyer and K.W. Morton. (1967). *Difference Methods for Initial Value Problems*. New York: Wiley.

D.R. Scheffler. (1996). CTH hydrocode simulations of hemispherical and ogival nose tungsten alloy penetrators perforating finite aluminum targets, In V.S. Shin and J.A. Zukas, eds, *Structures Under Extreme Loading Conditions*, PVP-vol. 325. New York: ASME, pp. 125-136.

D.R. Scheffler and L.S. Magness Jr. (1998). Target strength effects on the predicted threshold velocity for hemi- and ogival-nose penetrators perforating finite aluminum targets, In N. Jones, D.G. Talaslidis, C.A. Brebbia and G.D. Manolis, eds, *Structures Under Shock and Impact V (SUSI98)*. Southampton: Computational Mechanics Publications, pp. 285-298.

D.R. Scheffler and J.A. Zukas. (2000). Practical aspects of numerical simulation of dynamic events: material interfaces. *Int. J. Impact Engng*, 24, 821-842. also Army Research Laboratory, Report ARL-TR-2302, September 2002.

S.J. Schraml and K.D. Kimsey. (1998). Smoothed particle hydrodynamics simulation of disk-shaped penetrator impact, Army Research Laboratory, Report ARL-TR-1766, August.

W. Shyy, H.S. Udaykumar, M.M. Rao and R.W. Smith. (1996). *Computational Fluid Dynamics with Moving Boundaries*. Philadelphia: Taylor & Francis.

S.A. Silling. (1994). CTH reference manual: boundary layer algorithm for sliding interfaces in two dimensions. Sandia National Laboratories, Report SAND93-2487.

D.W. Stillman, L. Lum, J.O. Hallquist and D.J. Benson. (1993). *An arbitrary Lagrangian–Eulerian capability for LS-DYNA3D*, In E.P. Chen and V.K. Lik, eds, *Advances in Numerical Simulation Techniques for Penetration and Perforation of Solids*, AMD-vol. 171. New York: ASME, pp. 67-74.

J. Thomas and S. Jones. (1992). Numerically stable implementation of 3-D interface reconstruction algorithm, Report K-92-11U(R).

J.D. Walker and C.E. Anderson Jr. (1994). Multi-material velocities for mixed cells. In S.C. Schmidt, J.W. Shaner, G.A. Samara and M. Ross, eds, *High Pressure Science and Technology—1993*. Woodbury, New York: American Institute of Physics.

D.L. Youngs. (1987). An interface tracking method for a 3-D Eulerian hydrodynamics code, UK Atomic Weapons Establishment, Report AWRE 44/92/35.

J.A. Zukas. (1987). Fracture with stress waves. In T.Z. Blazynski, ed, *Materials at High Strain Rates*. London: Elsevier Applied Science Publishers.

J.A. Zukas (ed). (1990). *High Velocity Impact Dynamics*. New York: Wiley.

J.A. Zukas and S.B. Segletes. (1991). Hypervelocity impact on space structures, In T.L. Geers and Y.S. Shin, eds, *Dynamic Response of Structures to High-Energy Excitations*, AMD-vol. 127/PVP-vol. 225. New York: ASME, pp. 65-71.

J.A. Zukas, S.B. Segletes and J.R. Furlong. (1991). Numerical simulation and animation of impact phenomena. *Sci. Comput. Automat.*, 7(6), 15-29.

J.A. Zukas, J.R. Furlong and S.B. Segletes. (1992). Hydrocodes support visualization of shock-wave phenomena. *Comput. Phys.*, March/April, 1992.

Chapter 7
Experimental Methods for Material Behavior at High Strain Rates

7.1 Introduction.

It is necessary to remember that under conditions of high-rate loading, such as blast, impact and explosive loading, the assumptions inherent in the classical disciplines involving rigid-body mechanics no longer apply. Under dynamic loading conditions, inertia effects must be considered. Pressures can be generated which exceed material strengths by orders of magnitude. Since steady-state conditions no longer apply, the effects of stress wave propagation must be taken into account. This applies not only to the constitutive models used to describe high rate material behavior but also to the determination of the "material constants" which are used in the models. Handbook data, almost all of which is determined under low pressure and quasi-static conditions, is useless in dynamic situations save for first-cut approximations.

Here we run into a quandary. In order to determine material properties for very high strain rates, it is necessary to consider the details of inelastic wave propagation in a material. Quoting Nicholas [1982], "...There is, however, no direct method for obtaining dynamic property information because one must use wave propagation theory to study the phenomena which, in turn, requires a constitutive equation. But it is the constitutive equation that one seeks in the first place. Only an indirect or iterative procedure can be used, and the results cannot be uniquely established." The procedure recommended by Nicholas is the following:

(1) Assume the form of the constitutive equation(s) and the material constants in the equation.

(2) Perform a wave propagation experiment and establish or record the input or boundary conditions as well as information on the wave propagation phenomena such as strain–time profiles, free surface velocity measurements, wave velocities and so on.

(3) Make analytical predictions of observed phenomena from the assumed wave theory and constitutive relation.

(4) Compare predictions with experimental observations and revise the constants if necessary. Repeat steps 1–4.

252 *Introduction to Hydrocodes*

(5) Try another experimental configuration or other experimental observations using the same equation and constants to verify the predictions.

As can be seen, this is a trial-and-error procedure *which may not lead to a unique solution* (my emphasis). An important aspect of this procedure is to start with wave propagation theory and form of constitutive equation with a sound, physical basis.

In order to use high-strain-rate data in computations, the code user must know how the data were generated, as well as its limitations. Just as quasi-static data are of very limited use for high-rate applications, shear stress–strain data at nominal strain rates of $100\ s^{-1}$ are of limited use when studying spallation under hypervelocity impact, a condition of tensile failure of material at much higher strain rates. What experimental techniques are available for determining material properties under high strain rate conditions? These include [Nicholas, Rajendran & Sierakowski, 1998]:

- The split-Hopkinson pressure bar (SHPB), also known in the literature as the Kolsky apparatus. It comes in a number of configurations for determining material states in tension, torsion and shear.

- The Taylor cylinder test.

- The expanding ring test.

- Plate impact tests.

- Pressure–shear experiments.

Other techniques such as Charpy and Izod impact tests, drop towers, the Charpy pendulum test, rotating cam devices and accelerating mass devices exist as well. These will not be covered here. Refer instead to Meyers [1994], Nicholas [1982], Nicholas & Rajendran [1990] and other sources. Numerous configurations of these devices are used. They do not produce stress–strain data. They can provide some indication of bending or breaking strength. Strain rates achievable with such devices are generally below $1000\ s^{-1}$. Stress measurement is tricky. Often strain values must be assumed and the presence of stress wave ringing makes interpretation of the data limited.

7.2 The split-Hopkinson pressure bar (Kolsky apparatus).

The SHPB is the most widely used apparatus for determining dynamic stress–strain behavior of a wide variety of materials for strain rates in the range of $100–10{,}000\ s^{-1}$.

Experimental Methods for Material Behavior at High Strain Rates   253

The technique has been widely studied and the underlying assumptions thoroughly investigated (see Lindholm [1964], Bertholf & Karnes [1975], Nicholas [1982], Meyers [1994], among many]. Both the experimental setup and analysis are simple. A complete stress–strain curve is produced in the plastic region. No information is obtained in the elastic region. Where complete elasto-plastic curves are depicted, the elastic curve is typically obtained under static conditions and blended with SHPB results. Strain rates quoted with stress–strain curves obtained by using the SHPB are nominal, as the strain rate is not constant during the test.

A schematic of the SHPB is shown in Fig. 7.1. Figure 7.2 shows the SHPB configured for shear, tension and compression tests. A photograph of an actual SHPB setup is shown in Fig. 7.3.

A specimen is inserted between the incident and transmitter bars, Fig. 7.1. The incident and transmitter bars are chosen so that they remain in the elastic state during the experiment. The specimen undergoes elasto-plastic deformation.

Figure 7.4 is a schematic of the specimen and the strain pulses. As the incident bar is hit by the striker bar, an elastic pulse will propagate down the incident bar with a wave speed of $c = \sqrt{E/\rho}$. In the terms below, the subscripts i, r, t refer to incident, reflected and transmitted pulses in the various bars. Once the wave reaches the specimen, part of it will be reflected back along the incident bar because of the impedance mismatch between bar and specimen. Part of the wave is transmitted through the specimen. At the other end, part of the wave in the specimen will be transmitted into

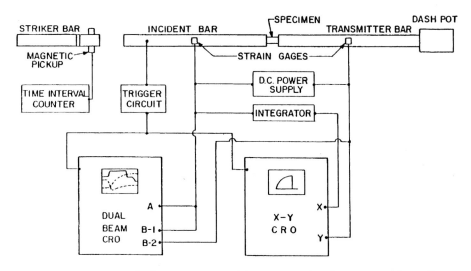

Figure 7.1 Schematic for apparatus and instrumentation for compression split-Hopkinson bar [Nicholas et al., 1998].

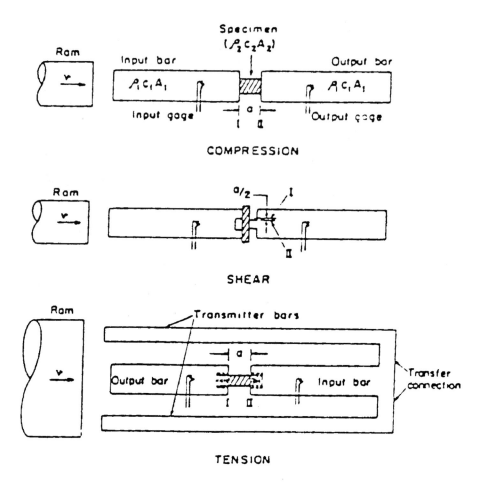

Figure 7.2 Split-Hopkinson bar arrangements [Nicholas *et al.*, 1998].

the transmitter bar and part reflected back into the specimen. The specimen therefore undergoes numerous wave reverberations. After a number of these, say 10–12 in metals for the sake of specificity, we fervently hope that a steady-state condition has been set up in the specimen. If we also take care to make the incident and transmitter bars sufficiently long so that reflected waves from the ends do not arrive until after we have made the necessary measurements, then we can determine the displacements at the specimen ends from:

$$u_1 = \int_0^t c_0 \varepsilon_1 dt, \tag{7.1}$$

$$u_2 = \int_0^t c_0 \varepsilon_2 dt, \tag{7.2}$$

Experimental Methods for Material Behavior at High Strain Rates 255

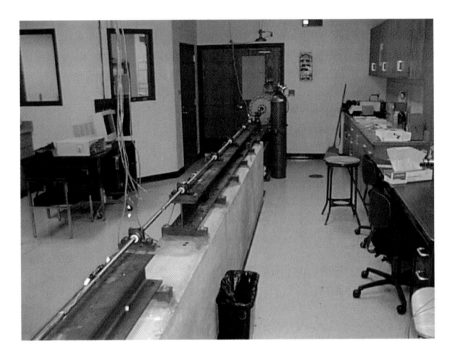

Figure 7.3 SHPB arrangement at the University of Delaware
[Courtesy: Prof. J.R. Vinson].

$$u_1 = c_0 \int_0^t (\varepsilon_i - \varepsilon_r) dt, \tag{7.3}$$

$$u_2 = c_0 \int_0^t \varepsilon_r dt. \tag{7.4}$$

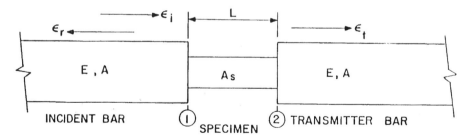

ε_i = INCIDENT PULSE
ε_r = REFLECTED PULSE
ε_t = TRANSMITTED PULSE
L = GAGE LENGTH

Figure 7.4 Positioning of specimen in the SHPB experiment [Nicholas et al., 1998].

The strain in the specimen is given by:

$$\varepsilon_s = \frac{u_1 - u_2}{L}, \tag{7.5}$$

$$\varepsilon_s = \frac{c_0}{L} \int (\varepsilon_i - \varepsilon_r - \varepsilon_t) dt. \tag{7.6}$$

Two assumptions are involved here:

- The pulses traveling in the incident and transmitter bars are *not* dispersed.
- The specimen is short enough so that a uniform state of stress exists in it due to multiple wave reverberations.

With the displacements determined, the strain in the specimen is given by:

$$\varepsilon_s = \frac{u_1 - u_2}{L}, \tag{7.7}$$

$$\varepsilon_s = \frac{c_0}{L} \int (\varepsilon_i - \varepsilon_r - \varepsilon_t) dt. \tag{7.8}$$

The forces acting on each end of the specimen are then:

$$P_1 = EA(\varepsilon_i + \varepsilon_r), \tag{7.9}$$

$$P_2 = EA\varepsilon_2, \tag{7.10}$$

$$P_{AVG} = \frac{EA}{2}(\varepsilon_i + \varepsilon_r + \varepsilon_t). \tag{7.11}$$

To go further we need to make some assumptions. We *assume* that the pulses travel down the bars essentially undispersed. We further assume that the specimen is short enough so that very many wave reflections will occur during the time for observation. If this indeed happens, then we take

$$P_1 = P_2, \tag{7.12}$$

so that

$$\varepsilon_i + \varepsilon_r = \varepsilon_t. \tag{7.13}$$

Experimental Methods for Material Behavior at High Strain Rates 257

With these results, the strain, stress and strain rate *in the specimen* may now be written as

$$\varepsilon_s = \frac{-2c_0}{L} \int_0^t \varepsilon_r \, dt, \tag{7.14}$$

$$\sigma_s = E \frac{A}{A_s} \varepsilon_t, \tag{7.15}$$

$$\dot{\varepsilon}_s = \frac{-2c_0}{L} \varepsilon_r. \tag{7.16}$$

A certain amount of ringing occurs in the test. As shown in Fig. 7.5, it is customary to fit a smooth curve through such regions.

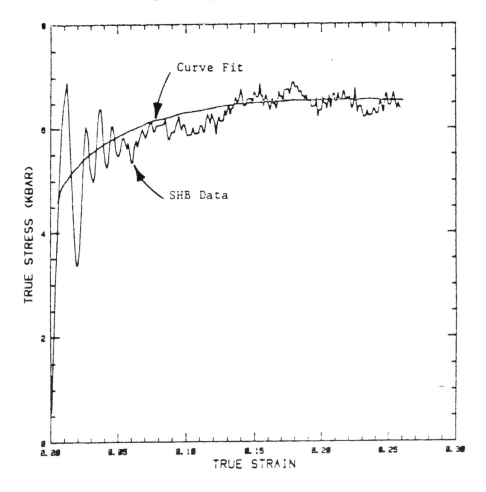

Figure 7.5 Curve fit and original data for a dynamic tesile SHPB test on 1020 steel [Nicholas *et al.*, 1998].

258 *Introduction to Hydrocodes*

Figure 7.6 shows typical Hopkinson bar data for a titanium alloy. Note that the strain rate in the experiment is not constant. It is therefore customary to quote an average strain rate when depicting the data.

A finite window exists for making the measurements. Figure 7.7 [Nicholas, 1980] shows a Lagrangian diagram for a tension Hopkinson bar. The tensile bar has a collar over the specimen. When the compressive pulse arrives from the incident bar the collar transmits the wave to the transmitter bar with the specimen remaining unloaded. At the end of the transmitter bar the compressive pulse is converted to a tensile wave in order to satisfy the boundary conditions at that end. When it arrives at the specimen, the collar falls away and the specimen is stressed in tension. Note from Fig. 7.7 that, at impact, stress pulses are propagated in both the striker bar and the incident bar. As the striker bar is short compared to the incident and transmitter bars, the stress pulse in it will reflect from its end, traverse the striker bar again and be passed, in part, into the transmitter bar. In the meantime, the original pulse in the incident bar passes through the collar into the transmitter bar, is reflected in tension and proceeds to the sample, as noted above. On its heels is the pulse form the striker bar. Thus, only a finite time exists to measure the incident, transmitted and reflected strains before the data is overwhelmed with noise form the wave reflections from the bar ends.

The assumptions inherent in the analysis of the SHPB become less and less valid as attempts are made to increase the strain rate. The apparatus has been used in tension

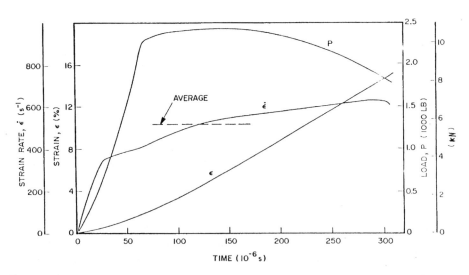

Figure 7.6 Typical tensile Hopkinson bar data for Ti 6Al-4V specimen [Nicholas et al., 1998].

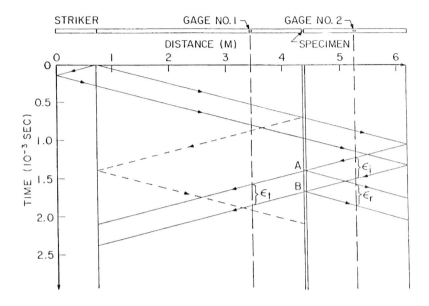

Figure 7.7 Lagrangian diagram for tensile SHPB test [Nicholas, 1980].

[Nicholas, 1981] and torsion [Frantz & Duffy, 1972] as well as compression. A large number of studies have been conducted to determine the valid operating ranges of the SHPB. These are summarized by Nicholas [1982]. Additionally, Follansbee & Frantz [1983] studied 3D dispersion effects in the bars and proposed corrections for the usual 1D analysis.

In addition to determining stress–strain data for metals, the SHPB has been used to test composite materials. Comprehensive reviews of dynamic testing of composite materials are given by Sierakowski [1988, 1997].

The SHPB has been used to determine failure strengths of brittle materials such as concrete and mortar in both tension and compression. Tests at elevated temperatures have also been performed by placing an oven over the specimen and heating it to the desired temperature. See Nicholas & Rajendran [1990] for descriptions and references to the relevant literature.

Figure 7.8 shows typical results obtainable with the SHPB (the topmost curve) in this case for AISI 304 steel. When combined with data obtained at other strain rates, a good appreciation is obtained of the strain-rate sensitivity of the material. Other examples which include effects of temperature for metals tested in tension are shown in Fig. 7.9. High-rate data are frequently published in the *DYMAT Journal* which can now be accessed through the internet.

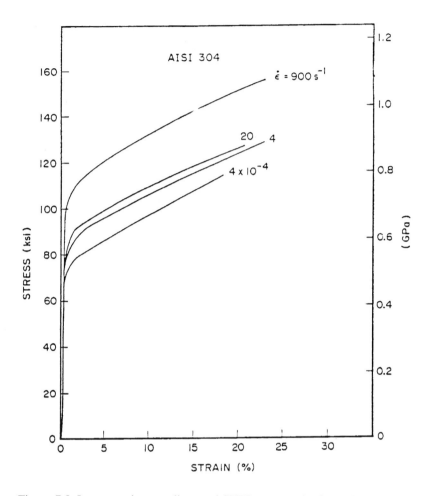

Figure 7.8 Low rate, intermediate and SHPB test results for AISI 304 steel [Nicholas et al., 1998].

7.3 The Taylor cylinder.

The Taylor cylinder test is a very valuable tool for obtaining a quick estimate of an *average* dynamic stress at an average strain rate of 100 s^{-1}. The test is easily performed. A schematic of two equivalent test conditions is shown in Fig. 7.10. A short specimen rod with a length-to-diameter (*L/D*) ratio of 3–5 is fired into a "rigid" anvil. The specimen is recovered and its total length, undeformed length and deformed diameter are measured, Fig. 7.11. From a simple analysis first developed by Taylor [1948], it is then possible to obtain a dynamic yield stress. Strain rates throughout the test are unknown. Both strains and strain rates are not uniform. A state of uniaxial stress is assumed in the analysis.

Experimental Methods for Material Behavior at High Strain Rates 261

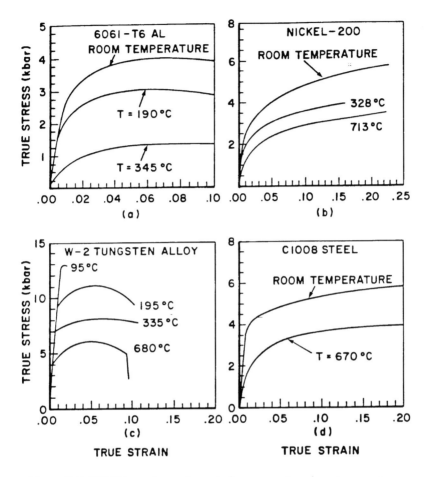

Figure 7.9 SHPB true stress–true strain curves at various temperatures
[Nicholas *et al.*, 1998].

Also, planar impacts are difficult to achieve in practice and, when dealing with strong materials, rigid targets are difficult to find. The test's two main virtues are that it is quick and easy to perform and it produces a useful estimate of a dynamic yield stress. An excellent discussion of the Taylor [1948] and Hawkyard [1969] analyses of the test technique can be found in Johnson [1970].

Refer to Fig. 7.12. Following Hohler & Stilp [1980], "…A rod of length L striking a target at time $t = 0$ produces a pressure jump of a certain level at the interface between rod and target and an elastic and plastic compression wave start to move into the rod. The elastic wave is reflected at the free rod end as a tensile wave and comes back to the plastic front that, in the meantime, moved only a small distance because of its lower wave speed c_{pl}. Now this elastic tensile wave is reflected again at the plastic front as a compression

262 *Introduction to Hydrocodes*

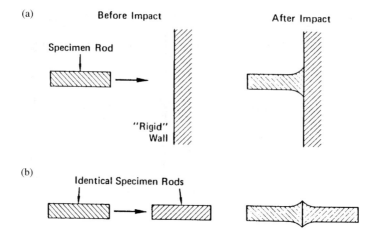

Figure 7.10 Schematic of (a) conventional Taylor test and (b) Alternative symmetric rod impact test [Nicholas et al., 1998].

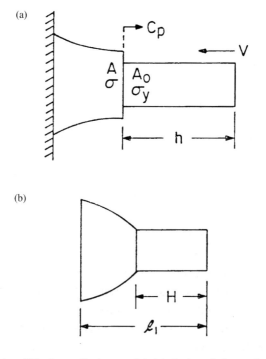

Figure 7.11 Schematic of Taylor cylinder model (a) during deformation and (b) after deformation [Nicholas et al., 1998].

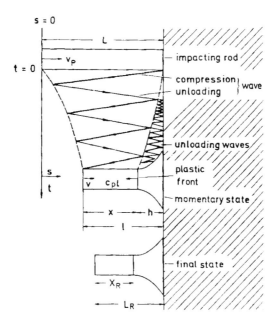

Figure 7.12 Elasto-plastic wave propagation and material deformation of rod during impact against a rigid target [Hohler & Stilp, 1980].

wave, moves back, and so on. Furthermore, there are rarefaction waves in the plastically deformed region starting when the elastic wave is reflected and moving to the right and to the left. This process results in a rod in which, at any moment, the plastic wave has moved a distance h. The undeformed rod length at this moment is x. The total length is $l = x + h$. The corresponding quantities of the final state are X_R and L_R." The resultant analysis produces a formula in terms of the residual undeformed length L, the total residual length, the projectile velocity v_P, the plastic wave speed c_{pl} and the projectile density ρ_p and yield strength σ_y, which can be solved for the yield strength:

$$\frac{X_R}{L} = \exp\left[-\left(\frac{v_p^2}{2} + v_p c_{pl}\right)\frac{\rho_p}{\sigma_y}\right]. \tag{7.17}$$

Hohler & Stilp [1980] note that there are two problems with the use of this formula. First, the elasto-plastic boundary is determined only with a great deal of difficulty form the residual rod geometry. Small errors in measurement significantly affect the determination of yield strength. Second, the plastic wave velocity is assumed to be constant, which is an approximation at best.

Some typical experimental results [Recht, 1990] are shown in Fig. 7.13. Whiffin [1948] performed a series of experiments at varying velocities. At the lower velocities,

264 *Introduction to Hydrocodes*

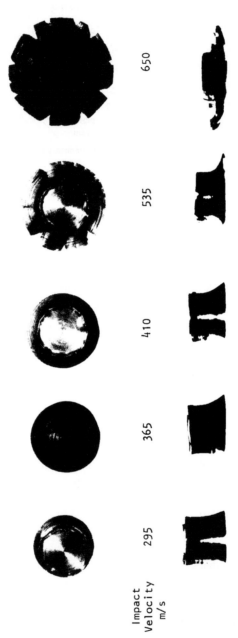

Figure 7.13 Deformation of cylinders after impact against hardened steel targets [Recht, 1990].

Experimental Methods for Material Behavior at High Strain Rates 265

the agreement between experimentally determined rod profiles and those determined by Taylor's analysis was quite good. At higher velocities significant deviations occurred. Some years later Hawkyard [1969] (see also Johnson [1970]) repeated the analysis. Whereas Taylor had assumed momentum conservation, Hawkyard based his analysis on energy conservation and obtained the following relations. The final deformed length of the cylinder is obtained from:

$$h = L(1 - e_0)\ln\left[\frac{1}{(1 - e_0)}\right], \qquad (7.18)$$

where e_0 is obtained from the expression:

$$\frac{1}{2}\rho_p v_p^2 = \sigma_y\left[\ln\left(\frac{1}{1 - e_0}\right) - e_0\right]. \qquad (7.19)$$

Although more cumbersome than Taylor's formula, it successfully determines the experimental profiles determined by Whiffin.

Wilkins & Guinan [1973] proposed a model which assumed deceleration of only the elastic part of the rod. A number of computational studies were performed. They indicated that inside the rod the plastic front position is nearer the rigid boundary that measurements of recovered samples would indicate. The simulations also indicated that the plastic wave speed is not constant. The plastic front jumps during the early stages of impact to an approximately fixed distance h from the rigid boundary. This results in the following equation:

$$\frac{L_R}{L} = \left(1 - \frac{h}{L}\right)\exp\left(-\frac{\rho_p v_p^2}{2\sigma_y}\right) + \frac{h}{L}. \qquad (7.20)$$

Here, h was found to be independent of v_p and proportional to L with

$$h/L = 0.12 = \text{const.} \qquad (7.21)$$

7.4 The expanding ring.

This experiment involves the rapid expansion of a circular ring which subjects the material to very high strain rates under a nominal uniaxial tension stress state in the hoop stress direction. See Nicholas [1982] for a description of the experiment and a history of its development.

266 *Introduction to Hydrocodes*

The geometry of the test is shown in Fig. 7.14. The experimental configuration is shown in Fig. 7.15. The analysis is shown in Fig. 7.16. It is based on the equation of motion of a segment of the ring, which is free of any external driving forces. The only forces acting are due to the hoop stress, which decelerate the ring. Typical results obtained from this experiment are shown in Fig. 7.17 [Nicholas et al., 1998].

While strain rates of up to 10,000 s^{-1} are achievable in the test, the strain rate decreases throughout the test. The test itself is very difficult to perform and requires very precise displacement measurements. These must then be differentiated twice in order to calculate stress. Attempts have been made to measure velocities in order to achieve greater accuracy [Warnes, Duffey, Karpp & Carden, 1981]. Another problem encountered is the inability to reproduce data from other tests or from different size rings. The reasons for this are discussed by Nicholas & Rajendran [1990].

The effective uses of this experiment are given by Nicholas & Rajendran [1990]: "…the expanding ring test technique is useful for the study of

- strain rate and inertia effects on the strain to failure,
- fragmentation, and
- constitutive model development.

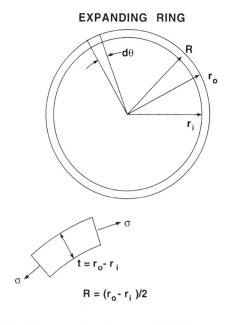

Figure 7.14 The expanding ring test [Nicholas et al., 1998].

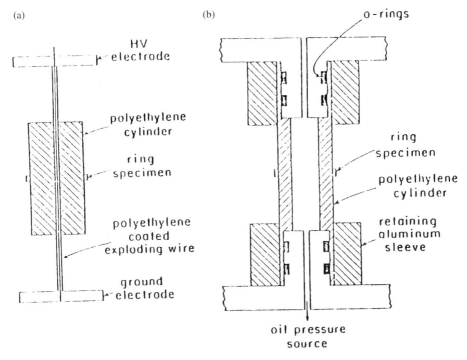

Figure 7.15 Configuration for expanding ring experiment: (a) dynamic and (b) static [Nicholas et al., 1998].

EXPANDING RING EQUATIONS

Sum of forces in radial direction:

$$F_r = -2\sigma t \sin(d\theta/2)$$

Mass $= \rho t R\, d\theta$

Conservation of momentum:

$$-2\sigma t \sin(d\theta/2) = \rho t R\, d\theta\, \ddot{R}$$

$$\boxed{\sigma = -\rho R \ddot{R}}$$

Assumes:
- Thin ring, $t \ll R$
- Uniaxial stress in ring
- Small strains, $\varepsilon = \Delta l / l$
- Incompressible (no Poisson's ratio effects)

Figure 7.16 Governing equations for the expanding ring test [Nicholas et al., 1998].

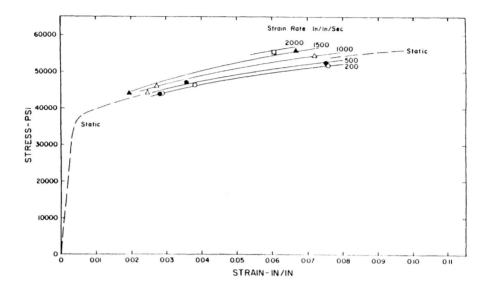

Figure 7.17 Dynamic stress–strain curves for 6061-T6 aluminum from expanding ring test [Nicholas *et al.*, 1998].

The velocity–time history measured in the expanding ring test can be best used to calibrate viscoplastic constitutive models rather than for obtaining direct information on stress–strain behavior."

7.5 Plate impact experiments.

Plate impact experiments have two main functions:

- To determine the Hugoniot for a given material at a given density. If you need to refresh your memory, see Chapter 3 and the outstanding and lucid description by Cooper [1996]. Even though you sometimes see references in the literature to the Hugoniot as being an equation of state, this is incorrect. The Hugoniot is the locus of all states that may be reached under a *shock transition*. By itself it is not a complete equation of state in that it does not account for energy states that can be reached by other than shock loading conditions. It is also an entirely empirical equation, good only for one material at a specific density. In order to extend its utility an energy term is added, as in the Mie–Gruneisen equation (see Chapter 3) to reach other states near the Hugoniot. In the literature this is called an "off-Hugoniot"

Experimental Methods for Material Behavior at High Strain Rates   269

modification. Of course, with simple energy term tacked onto polynomial Hugoniot fits one does not get much in the way of generality. However, in many applications this is good enough for all practical purposes. Remember, Hugoniots are more practical than theoretical. I once took a class in solid mechanics taught by a professor who was as rigorous and thorough as he was wise. In the middle of a complex derivation he jumped from one stage to another without providing any motivation for the approach. Given his reputation I assumed that I had missed something obvious. Finally someone got up the courage to ask. The professor sheepishly admitted that there was no good reason for doing what he did. The motivation for it was that "...we do it because it works." A true engineer!

- To determine the spall strength of a material.

The experimental setup, Fig. 7.18, requires some care but is fairly simple. A plate, called the "flyer," is launched either explosively or from a gas gun against another stationary plate, referred to as the "target." Shock waves propagate into both flyer and target. As in the SHPB, there is a finite window during which observations can be made.

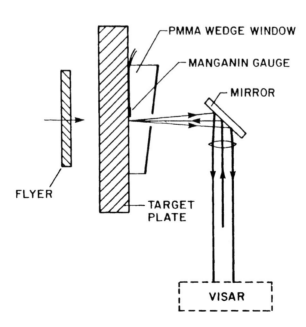

Figure 7.18 Schematic of plate impact experiment [Nicholas et al., 1998].

That time is governed by the thickness and width of the flyer plate. After the initial shock wave transits, it is reflected from the flyer rear surface as a rarefaction wave and proceeds into the target. The effective observation time is the period between arrival of the original compressive pulse in the target from the impact interface and the arrival of the rarefaction wave from the flyer. This of course assumes that the lateral dimensions of the plates are large enough so measurements are completed before signals arrive from the lateral boundaries and destroy the uniaxial strain state which exists at the time measurements of target rear surface deflections are made. The plate impact experiment is described by Karnes [1968].

Focus on the target plate. After impact, a compressive shock moves into the target plate. It reflects at the rear surface as a tensile wave and moves toward the impact interface of the target plate. After the tensile shock has traveled some distance from the rear of the target, failure by spallation may occur if the proper conditions are met. The physics of the situation for a variety of stress pulse shapes is discussed by Rinehart [1960]. At first, it was thought that a critical level of tensile stress was responsible for spallation. There are tables in Rinehart's book that give values for the critical tensile stress for a number of materials. Since then, sufficient evidence has accumulated to suggest that a realistic spall criterion cannot depend on a single mechanism. Among the most interesting experiments, which indicate that the duration of the tensile pulse at the spall plane plays a significant role in spallation, are the experiments of Blinkow and Keller, as discussed in Oscarson & Graff [1968]. They observed that the flyer velocity required to cause incipient spall decreases with increasing flyer thickness in an approximately linear manner. Since impact velocity is directly related to stress and flyer thickness can be related to pulse duration, a definite time dependence of spall is implied. Other factors such as the

- effects of precompression at the spall plane,
- strain and stress rates at the spall plane,
- spatial stress gradients,
- temperature,
- material strength, and
- metallurgical characteristics

have been considered [Oscarson & Graff, 1968]. Meyers & Aimone [1983] performed an exhaustive study of spallation. A good discussion of the subject can be found in Meyers [1994]. An example of spallation can be seen in Fig. 7.19.

Experimental Methods for Material Behavior at High Strain Rates 271

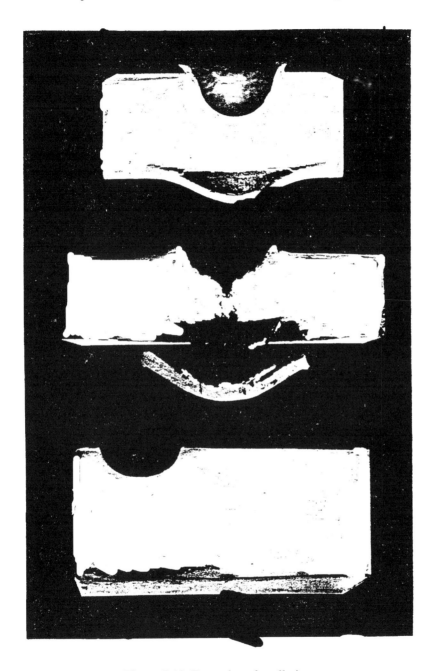

Figure 7.19 Examples of spallation.

7.6 Pressure–shear experiments.

The pressure–shear experiment was conceived and developed by Prof. Clinton and his co-workers at Brown University. The intent is to subject materials to states of combined compression and shear at very high rates. The oblique impact experiment [Abou-Sayed *et al.*, 1976; Abou-Sayed and Clifton, 1977] involves impacting a flyer plate against a target as in the conventional plate experiment except that the impact occurs at obliquity (Fig. 7.20). The impact generates both a compression wave and a shear wave. The normal and transverse velocities on the rear surface are recorded with a pair of velocity interferometers [Kim & Clifton, 1980] (Fig. 7.21). There are two versions: Fig. 7.20(a), the original version, is a pressure and shear wave propagation experiment in which wave motion is studied using an assumed constitutive equation. The second configuration, Fig. 7.20(b), uses the concept of the SHPB by sandwiching a soft specimen material between high impedance hard plates which remain elastic. In this configuration the specimen achieves a combined state of high compression in uniaxial strain while simultaneously being subjected to high shear strain rates, Fig. 7.22. Figure 7.23 has representative dynamic stress–strain curves for OHFC copper (see also Fig. 7.24).

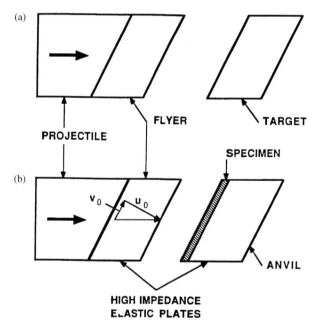

Figure 7.20 Schematic of pressure–shear experiment showing (a) compression–shear wave propagation testing and (b) high strain rate shear test configurations [Nicholas *et al.*, 1998].

Experimental Methods for Material Behavior at High Strain Rates 273

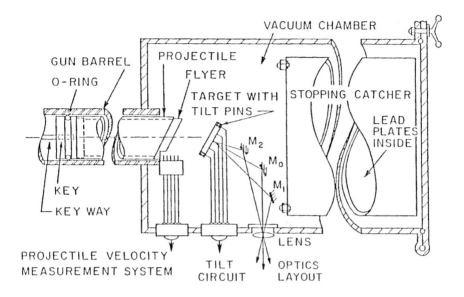

Figure 7.21 Schematic of pressure–shear experiment [Kim & Clifton, 1980].

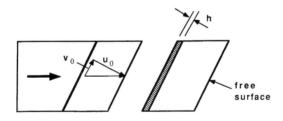

NORMAL AND SHEAR STRAIN RATES IN SPECIMEN:

$$\dot{\varepsilon} = \frac{u_0 - u_{fs}}{h}$$

$$\dot{\gamma} = \frac{v_0 - v_{fs}}{h}$$

NORMAL AND SHEAR TRACTIONS ON SPECIMEN:

$$\sigma = \frac{1}{2}\rho c_1 u_{fs} \quad c_1 = \text{longitudinal wave speed}$$

$$\tau = \frac{1}{2}\rho c_2 u_{fs} \quad c_2 = \text{shear wave speed}$$

Fig. 7.22 Pressure–shear experiment [Nicholas *et al.*, 1998].

274 *Introduction to Hydrocodes*

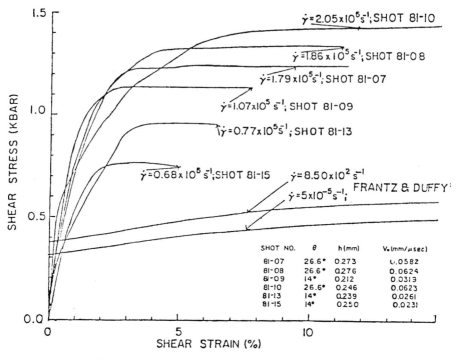

Figure 7.23 Dynamic stress–strain curves for OFHC copper [Li, 1982].

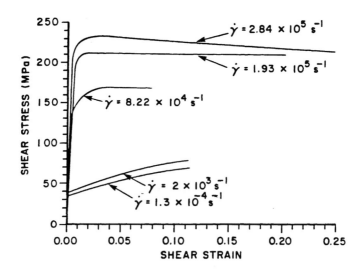

Figure 7.24 Dynamic plastic response of copper [Huang and Clifton, 1985].

7.7 Summary.

Other techniques exist for measuring various aspects of the dynamic behavior of materials. The above have been singled out since considerable data from these experiments can be found in the literature. This is especially true of the split-Hopkinson bar, which was a laboratory curiosity in the 1950s and has now become a standard tool in university, government and industrial materials research laboratories. To use high-rate data effectively in computations it is necessary to know how it was obtained as well as the limitations of the experimental method so that the information not be misused. The ultimate responsibility for a computational result rests with the scientist or engineer performing the calculation. In response to queries about the constitutive model used and sources of data for the material constants of the model, such statements as "…I don't know much about this area so I asked my buddy down the hall to give me some numbers" are music to the ears of litigators investigating disasters that involved computational efforts.

The purpose of the above thumbnail descriptions is to give current and future code users and idea of the methods currently available for obtaining high-strain-rate data, some of their advantages and drawbacks and some key references where additional information may be found. A strong familiarity with this topic is necessary for success in computational efforts. The biggest single limitation on today's computer codes remains the primitive state on the modeling of high rate behavior, *especially in so far as the modeling of failure is concerned*. With data from the above techniques, very decent predictions of large deformations can be made, especially in situations dominated by inertia. Going beyond this, however, to predict local variables such as stress, strain, the onset of failure, requires both improved constitutive models and, going hand-in-hand, straightforward experiments for determining the free parameters in those models. Multi-dimensional, multi-parameter models for dynamic behavior may be theoretically elegant, but if they include so many free parameters that up to a year's time is required to determine them, they lack practicality. Better a simpler approach that lends itself to back-of-the-envelope estimates coupled with laboratory tests and simple material characterization to get a timely answer than endless grinding with computer models where five or more parameters need to be guesstimated.

REFERENCES

A.S. Abou-Sayed and R.J. Clifton. (1977). Analysis of combined pressure–shear waves in an elastic/viscoplastic material. *J. Appl. Mech., Trans. ASME*, 44, 79-84.

A.S. Abou-Sayed, R.J. Clifton and L. Hermann. (1976). The oblique-plate impact experiment. *Exp. Mech.*, 16, 127-133.

L.D. Bertholf and C.H. Karnes. (1975) *J. Mech. Phys. Solids*, 23, 1.

P.W. Copper. (1996). *Explosives Engineering*, New York: Wiley-VCH.

P.S. Follansbee and C. Frantz. (1983). Wave propagation in the split Hopkinson pressure bar. *Trans. ASME, J. Engng. Mat. Tech.*, 105, 61-66.

R.A. Frantz and J. Duffy. (1972). The dynamic stress–strain behavior in torsion of 1100-0 aluminum subjected to a sharp increase in strain rate. *Trans. ASME, J. Appl. Mech.*, 39, 939-945.

J.B. Hawkyard. (1969). Mushrooming of flat-ended projectiles impinging on a flat, rigid anvil. *Int. J. Mech. Sci.*, 18, 379-385.

V. Hohler and A.J. Stilp. (1980). Study of the penetration behavior of rods for a wide range of target densities. *5th Intl. Symp. Ballistics, Toulouse, France*, Washington DC: ADPA.

S. Huang and R.J. Clifton. (1985). Dynamic plastic response of OFHC copper at high shear strain rates. In K. Kawata and J. Shioiri, eds., *IUTAM Symp. on Macro- and Micro-Mechanics of high velocty deformation and Fracture*, 63-74, Berlin: Springer-Verlag.

W. Johnson. (1970). *Impact Strength of Materials*. London: Edward Arnold.

C.H. Karnes. (1968). The plate impact configuration for determining mechanical properties of materials at high strain rates. In U.S. Lindholm, ed, *Mechanical Behavior of Materials Under Dynamic Loads*. New York: Springer-Verlag.

K.S. Kim and R.J. Clifton. (1980). Pressure–shear impact of 6061-T6 aluminum. *J. Appl. Mech., Trans. ASME*, 47, 11-16.

C.H. Li. (1982). A pressure–shear experiment for studying the dynamic plastic response of metals at shear strain rates of 10,000/s. Ph.D. Dissertation, Brown U., Providence, RI.

U.S. Lindholm. (1964). Some experiments with the split Hopkinson pressure bar. *J. Mech. Phys. Solids*, 12, 317-335.

M.A. Meyers. (1994). *Dynamic Behavior of Materials*. New York: Wiley.

M.A. Meyers and C.T. Aimone. (1983) *Prog. Mater. Sci.*, 28, 1.

T. Nicholas. (1980). Tensile testing of materials at high-rates of strain, USAF, Report AFWAL-TR-80-4053, Wright-Patterson AFB, OH.

T. Nicholas. (1981). Tensile testing of materials at high rates of strain. *Exp. Mech.*, 21, 177-185.

T. Nicholas. (1982). Material behavior at high strain rates. In J.A. Zukas, T. Nicholas, H.F. Swift, L.B. Greszczuk and D.R. Curran, eds, *Impact Dynamics*, New York: Wiley, Chapter 8.

T. Nicholas and A.M. Rajendran. (1990). Material characterization at high strain rates. In J.A. Zukas, ed, *High Velocity Impact Dynamics*, New York: Wiley, Chapter 3.

T. Nicholas, A.M. Rajendran and R.L. Sierakowski. (1998) *Material Behavior at High Strain Rates*. notes for the short course Material Behavior at High Strain Rates held in Monterey, CA on April 28–May 1.

J.H. Oscarson and K.F. Graff. (1968). Battelle Memorial Institute, Report BAT-197A4-3.

R.F. Recht. (1990). High velocity impact dynamics: analytical modeling and plate penetration dynamics. In J.A. Zukas, ed, *High Velocity Impact Dynamics*, New York: Wiley, Chapter 7.

J.S. Rinehart. (1960). *Stress Transients in Solids*. Santa Fe, NM: HyperDynamics.

R.L. Sierakowski. (1988). High strain rate testing of composites. In A.M. Rajendran and T. Nicholas, eds, *Dynamic Constitutive/Failure Models*, USAF, Report AFWAL-TR-88-4229. OH: Wright-Patterson AFB.

R.L. Sierakowski and S.K. Chaturvedi. (1997). *Dynamic Loading and Characterization of Fiber-Reinforced Composites*. New York: Wiley.

G. Taylor. (1948). The use of flat-ended projectiles for determining dynamic yield stress. I. Theoretical considerations. *Proc. R. Soc. Lond.*, A194, 289–299.

R.H. Warnes, T.A. Duffey, R.R. Karpp and A.E. Carden. (1981). An improved technique for determining dynamic material properties using the expanding ring. In M.A. Meyers and L.E. Murr, eds, *Shock Waves and High Strain Rate Phenomena in Metals*. New York: Plenum.

A.C. Whiffin. (1948). The use of flat-ended projectiles for determining dynamic yield stress. II. Tests on various metallic materials. *Proc. R. Soc. Lond.*, A194, 300.

M.L. Wilkins and M.W. Guinan. (1973). Impact of cylinders on a rigid boundary. *J. Appl. Phys.*, 44, 1200.

Chapter 8
Practical Aspects of Numerical Simulations of Dynamic Events

8.1   Introduction.

Most of the work done in the area of fast, transient loading is experimental in nature. This is either due to complexities of geometry or the nonlinearity of material behavior or both. Closed-form analytical solutions are generally rare and apply only to some small subset of the overall problem.

Numerical solutions, in the form of finite-difference and finite-element codes, have been successfully used in the past. In particular, the combination of experiments, numerical solutions, and dynamic material characterization has been shown to be very effective in reducing both manpower requirements and cost [Herrman, 1977]. However, the computer codes available for dynamic analyses are quite complex. Considerable experience with both the codes and the physical problems they are intended to solve is vital. Also critical is the determination of material constants for the various constitutive models available at strain rates appropriate to the problem.

Figure 8.1 shows the dramatic effect that these factors can have on computed results. The figure, provided by Dr. Paul Senseny, shows the results obtained by four different users of the DYNA code for a problem involving air-blast loading of a silo door. Each of the code users worked independently. Note that four people with four different backgrounds produced four distinctly different results *with the same code* in solving the same problem.

There are several reasons for such an occurrence. Remember that codes have a number of free parameters that must be set by users. These occur mainly in the subroutines dealing with material modeling. If a user has some preconception of what the answer should be, these parameters can be adjusted to produce something close to expectations. Recall also that for successful computations an appropriate constitutive model employing data developed from dynamic experiments at strain rates appropriate to the problem need be used. This is a major requirement. If it is not met, incorrect physics is being computed. Thus any correlation between calculations and experiments is purely fortuitous.

But there is more. For each calculation, the user must be sure to use an appropriate mesh. Too few cells/elements and the high-frequency components, which

280 *Introduction to Hydrocodes*

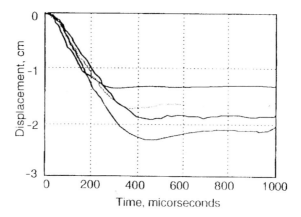

Figure 8.1 Four different answers from four different users all using the DYNA code (Courtesy: Dr. Paul Senseny).

dominate material failure, are filtered out; too many and the calculation is needlessly expensive. Once the mesh is chosen, an appropriate artificial viscosity must be selected. Too much viscosity damps the solution. Too little induces artificial ringing. The choice of cell or element type affects the outcome. Triangular elements work fine for static applications but are notorious for the problems they induce in dynamic calculations unless excess degrees of freedom are accounted for. Material failure modeling and characterization are in their infancy now; yet these factors strongly influence numerical solutions. Thus, one can understand why four different users with different backgrounds and different levels of mastery of the physics of the problem as well as the computational tool used can come up with four different solutions!

There exist many cases in computational continuum dynamics where tradeoffs must be made to achieve reasonable results in a reasonable time at finite costs. Szabo & Actis [1996] point out the importance of timely solutions in industry. Quoting from an engineer's experience at a major industrial laboratory, they point out that "...the stress engineer may not be able to guide the design because by the time he has generated his several thousand degree-of-freedom finite-element model, prototypes are already being made." Anyone can perform finely resolved 1D calculations almost by rote. Two-dimensional calculations, though now largely performed on workstations and personal computers, require a keen knowledge of the problem and numerical simulation methods. This holds even more for successful 3D simulations that, with all their compromises, will still be expensive and require significant computer resources [Zukas, 1989; Zukas & Scheffler, 2000].

8.2   Difficulties.

When a user acquires a code, he/she receives a package in which certain decisions have been made by the developer. The user has no control over some of the approximations that come as part of the package but must be aware of them, as they will affect all solutions that the user will generate with the package. This section briefly describes these errors and cites sources in the finite-difference/finite-element literature where extensive discussions can be found. These are problems inherent in transforming a physical problem into a discrete model and solving it on computers with limited precision through the medium of finite differences and finite elements. Each difference scheme and each element have unique characteristics that can affect the numerical solution. For example, most implicit schemes for advancing the solution in time have a certain amount of viscosity built into them over which the user generally has no control. Yet, the user must be aware of this characteristic when he/she sees his/her numerical solution drift out of phase with an experimental or analytical result. Certain finite elements are prone to locking (becoming excessively stiff) when the constitutive relationship is evaluated at element Gauss points. This can be overcome by evaluating the constitutive equations at one point (usually the element centroid). This procedure is called "under integration." It eliminates locking but gives rise to spurious deformation modes known as "hourglassing." A user, unaware of this, might interpret such numerical noise as a physical response. Explicit integration schemes are only conditionally stable. Using stability fractions built into explicit hydrocodes without consideration of the problem being solved can lead to unstable solutions that can degrade over hundreds of cycles and could be interpreted by novice code users as a physical response of the system being modeled.

The errors inherent in finite-element modeling are well discussed by Utku & Melosh [1983], as are guidelines for mesh preparation by Melosh & Utku [1983]. A superb work on practical aspects for static and dynamic finite-element analyses is the book by Meyer [1989]. Extremely valuable insights into finite-element design and its practical application are to be found in MacNeal's book [MacNeal, 1994]. Morris & Vignjevic [1997] review error control and error bounding methods for finite elements. They also present a method for error control in the idealization phase of a full finite-element analysis. These, together with personal experience, are the main references for the material in this section.

An analyst begins by examining some physical problem that needs to be modeled numerically. This may be necessary because it is too expensive, too difficult, or just plain impossible (as in nuclear testing) to perform parametric experiments. It might be because the information needed does not lend itself to direct measurement. It is often the case [Herrman, 1977; NMAB, 1980] that the combined use of experiments and calculations—including some material characterization at high strain rates—produces more information in less time and at less cost than reliance on experiments or calculations alone. At first glance, a problem may appear to be extremely complex. Pressure vessels come with various cut-outs, end caps that can range from circular plates through hemispherical shells, pipes coming in and out, some irregular sections, weld points, and a countable infinity of bolts and assorted restraining devices. Face it; real structures are messy to model numerically. They can be mounted to floors and walls and receive and transmit loads to these through a variety of mounts (shock isolation is popular in Japan and other earthquake-prone areas). The space shuttle has a nice, clean shape, but most space structures tend to show greater similarity to the highly irregular surfaces of the space cruisers in Star Wars movies than to the shuttle. Thus, high-velocity impact-generated debris and ricocheting projectiles can interact with critical, externally mounted equipment such as antennas and solar panels.

Three-dimensional impact problems are the rule rather than the exception. Uniform grid resolution is generally not possible in practical problems. Some gradation of the mesh is required. If this is not done with great care, spurious signals and assorted numerical artifacts arise in the calculation, leading to instabilities or a masking of the desired results. Even coarse calculations can require in excess of 40 central processing unit (CPU) hours on modern computers. If sufficient memory is not available to run the problem "in-core," then extensive buffering between main memory and mass storage is required, further increasing turnaround time and cost. For sufficiently large problems on a small central memory machine, it is possible to approach situations where the bulk of CPU time is spent on input–output operations and only a small fraction is spent in advancing the solution, rendering the computation uneconomical.

Does this mean that 3D calculations cannot be done today? Not at all; quite a few practical problems have been successfully addressed with 3D codes. However, compromises are required, and these, in turn, require a keen understanding of the physical problem, the effects of discretization on that problem, and the effects that numerical artifacts (such as uneven resolution in different coordinate directions, mixing of implicit and explicit integration schemes or explicit–explicit partitions, choice of mesh or element type, effects of sliding surfaces or interfaces, and the use of various viscosities to stabilize computations) have on the solution.

All the details and materials in a real structure usually cannot be accounted for in a numerical simulation. Hence, the analyst must now make a tractable model without sacrificing the essential elements that make up the response of the physical structure. The first order of business is to simplify the physical system by taking all essential geometric and material features that govern its response into account. Some simplification of geometry and lumping of masses is inevitable. Depending on the situation, one may be able to employ specialized mathematical models such as beam, plate, or shell theory (or a combination of these). Preliminary analyses need to be made to determine whether large strains and rotations constitute a part of the response or whether a linear analysis will suffice. A number of uncertainties are introduced in this process, some of which cannot be quantified. Other uncertainties are due to variabilities in material properties, loading, fabrication, and other factors. For example, rolled homogeneous armor (RHA) steel is used extensively in military construction. It can safely be said that RHA is rolled. It is also used as armor. However, the military specifications that govern the production of RHA have wide tolerances so that it is anything but homogeneous. Material properties (primarily hardness) are known to vary by as much as 10% within a lot of RHA and up to 30% from lot to lot. This makes single tests (the famous "one-shot statistics") useless and correlation between numerical results and experiments unlikely unless a statistically meaningful number of tests have been done. Simple go/no-go ballistic tests can cost upward of $2,000 each. Instrumented field tests can run from $10,000 to $100,000 each. As a rule, then, a statistically significant data set is almost never available.

8.3 Idealization.

The ultimate goal of this idealization process—the transition from a complex physical model to a simpler one but incorporating all the relevant physics—is a mathematical model consisting of a number of equations that closely represent the behavior of the physical model. A formal seven-step process was proposed by Morris & Vignjevic [1997]. An experienced analyst will go through the procedure guided by a few principles and much insight garnered over a long career. The time required might take weeks to months, depending on the complexity of the physical system and the accuracy required in the analysis. If the assumptions of the mathematical model are reasonable, very little mathematical modeling error results. If this is poorly done, say if the height-to-span ratio of beam theory is violated, the thickness-to-radius ratio for thin shells is not satisfied, a poor choice of material properties is made, or the constitutive model omits a critical item such as dynamic failure, serious errors can be incurred. Such synthesis is not taught in schools but learned in apprenticeship with an experienced modeler. Fortunately, in the vast majority of cases, this is done very well so that mathematical modeling errors are negligibly small, or at least smaller than errors committed by code users, the topic of the next section.

A solution is needed for the mathematical model. Since most problems involving high-strain-rate loading are not analytically tractable, recourse is made to a computer. In the process of computing, especially in matrix operations, situations occur where differences between numbers of almost equal magnitude must be taken. This can lead to situations where the roundoff error can completely overwhelm the computed quantity. Given a machine, there is always a limit to the mesh refinement beyond which computed quantities may be 100% erroneous. Roundoff errors may be kept negligibly small by using "...longer word length machines and double precision arithmetic with not too refined finite element meshes" [Utku & Melosh, 1983].

Most dynamic analyses for problems involving wave propagation are done with the simplest elements—constant or linear strain triangles and quadrilaterals in 2D computations, tetrahedra and hexahedra in 3D computations. Many early (mid-70s) finite-element codes, such as Lawrence Livermore National Laboratory's DYNA and IABG's DYSMAS-L, incorporated high-order elements in their initial formulations. With experience, these were dropped. Many transient response calculations in the wave propagation regime involve the presence of steep stress gradients and shock waves. It has been found with experience that there is marginal increase in accuracy but considerable increase in cost in trying to model what are essentially discontinuities with higher-order polynomials. The characteristics of these elements, and the problems of locking and reduced integration associated with them, are lucidly discussed in the book by MacNeal [1994]. The original literature and the code manuals should also be reviewed in order that these elements not be misused.

Code users should also be aware of the viscosity built into implicit temporal integrators and the conditional stability of explicit integrators. Additional information can be found in Belytschko & Hughes [1992], Donea [1978], and Zienkiewicz [1983].

8.4 The human factor.

It is generally accepted that, if the idealization from physical system to mathematical model is done well, the errors due to truncation, roundoff, and other properties of finite-element or finite-difference schemes can be readily detected and contribute to no more than about 5% of the total solution error. Solutions, however, can be totally invalidated by poor choice of mesh, failure to include relevant physics in the constitutive description, using a limited or inappropriate database to evaluate the constants of a constitutive model, failure to recognize instabilities, or the effects of contact surfaces on numerical solutions. In short, computational techniques come with certain built-in limitations that are easily recognizable and, with few exceptions, controllable. To really foul up requires a human.

Some, but hardly all, of the errors in the application of computer codes to practical problems involve the following:

Meshing: a code user has a wide choice of elements available in any commercial code. Having selected one or more, the user can then vary element aspect ratio, the arrangement of elements or element groups, choose uniform or variable meshing, and even introduce abrupt mesh changes. All of these will influence the solution to some degree.

Constitutive model: assuming an appropriate model has been selected to account for material behavior under high-rate loading, criteria for material failure, and descriptions of post-failure behavior, the problem of selecting values for the various material constants in the constitutive model remains. These must be selected from experiments conducted at strain rates appropriate to the problem. In many cases, data may not be available. This is particularly true for situations involving material failure. The estimation of these factors then depends on the skill, knowledge, and experience of the user, and computed results will vary accordingly.

Contact surfaces and material transport: Lagrangian codes incorporate a wide variety of algorithms to account for contact–impact situations. Eulerian codes have a variety of methods for determining the transport of material from one cell to another. Each algorithm uniquely affects the solution. Some codes have incorporated a variety of algorithms and allow the user a choice. The burden is then on the user to choose wisely, and this cannot be done without a knowledge of how the various algorithms affect the solution of both global (e.g., displacements) and local (e.g., strain) variables.

Shortcuts: because of the expense involved in 3D calculations, recourse is sometimes made to plane strain solutions. These are 2D approximations of 3D phenomena. Sometimes they produce excellent results, sometimes disasters.

8.5   Problems related to computational meshes.

Theoretically, the ideal mesh is uniform in all coordinate directions and "converged" for the critical variable for the problem; that is, it is small enough to give accurate results so that further refinement dramatically runs up the cost of computing with negligible improvement in accuracy. In practice, especially when performing 3D calculations, this goal is impossible to achieve. Compromises must be made, and the educated analyst must know what effect these compromises have on the numerical solution to his/her problem. In short he/she must know the physical problem and understand how compromises to the computational mesh affect the solution.

There are a number of factors that affect mesh integrity. These include element aspect ratio, element arrangement, uniform vs. graded meshes, and abrupt changes in meshes. Each is now looked at in turn.

a)   *Element aspect ratio.*

Ideally, calculations would be done with 1:1 aspect ratios in the elements. This hardly ever happens in practical 2D and 3D calculations. Thus, it would be nice to know

286 Introduction to Hydrocodes

how the solution is affected when elements with aspect ratios exceeding 1:1 are used. Creighton [1984], using the EPIC-2 code, looked at the effects of aspect ratio, artificial viscosity, and triangular element arrangement on elastic and elasto-plastic impact situations. EPIC uses constant-strain triangular elements that can be arranged in a number of ways. The possibilities included in Creighton's study are shown in Fig. 8.2. The elastic calculations were performed for a steel bar with length-to-diameter (L/D) ratio of 100 striking a rigid barrier at 3.048 m/s. Numerical solutions were compared with an exact solution by Skalak [1957], Fig. 8.3, which takes into account the effects of radial inertia. Calculations were performed with one (400 elements), two (1600 elements), and three (3600 elements) crossed triangles (four triangles per quadrilateral) across the radius of the bar (length, $L = 12.7$ cm; diameter, $D = 0.254$ cm). All calculations were done with an aspect ratio of 1:1. The axial force in the rod as a function of position was compared with Skalak's solution at various times.

As expected, the results show the grid acting as a frequency filter. As the grid is resolved, more and more ringing is computed behind the wavefront. Enough high-frequency components are accounted for in the 3600-element calculation to compare very favorably with the analytical solution.

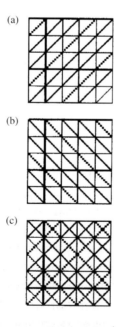

Figure 8.2 Element orientations in the EPIC-2 code. (a) IDIA = 1 Hypotenuse of the triangle is drawn from the lower left hand corner to the upper right hand corner. (b) IDIA = 2 Hypotenuse of the triangle is drawn from the upper left hand corner to the lower right hand corner. (c) Quadrilateral elements comprised of four triangles.

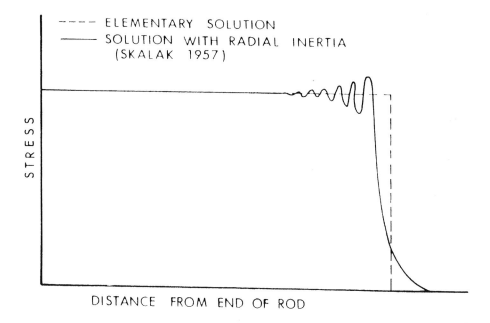

Figure 8.3 Comparison of elementary and Skalak solutions for bar impact (Skalak solution is drawn freehand) [Zukas, 1982].

Holding the number of elements across the bar radius at three, the element aspect ratio was now increased from 1:1 to 4:1. Again, the filtering characteristics of the mesh are clearly shown. High-frequency components are gradually suppressed until, with the 4:1 mesh, only the ringing directly behind the wave front remains (Fig. 8.4).

b) *Element arrangement.*

The arrangement or mixing of elements within a computational mesh can be a problem. Figure 8.2 shows three possible arrangements of triangular elements. This aspect of mesh generation was also investigated [Creighton, 1984]. The specific problem considered was the impact of a steel sphere with a 5.08-cm thick aluminum target at a velocity of 1524 m/s. The two-triangle orientations (Fig. 8.2(a) and (b)) gave rise to asymmetries in the calculation. Depending on the orientation of the diagonal, results were either too stiff or too soft. Optimal performance was achieved using the four-triangles-per-quadrilateral grouping (Fig. 8.2(c)). Similar results had been obtained by Zukas [1979]. The definitive study on element arrangement for accurate elasto-plastic solutions was done by Nagtegaal, Parks & Rice [1974].

Johnson and co-workers [1981, 1990] observed that, for 2D triangular elements, accuracy is much improved if the elements are arranged in a crossed-triangle (four

288 *Introduction to Hydrocodes*

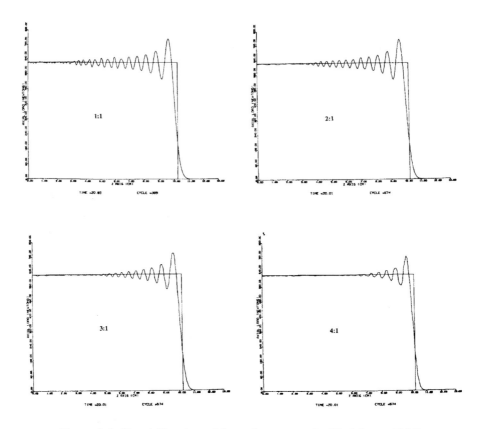

Figure 8.4 Signal filtration with mesh aspect ratio [Creighton, 1984].

triangles within a quadrilateral) arrangement or if a single average pressure is used for each group of two adjacent triangles (two triangles within a quadrilateral). In three dimensions, Johnson, Stryk, Holmquist & Souka [1990] performed a number of simulations using configurations of 6 tetrahedral elements within a brick, 6 tetrahedral elements per brick using a single average pressure for all 6 tetrahedra, as well as an arrangement of 24 tetrahedral elements per brick. This latter case, though the most costly configuration, was expected to produce the best results since the asymmetries present in the 6 tetrahedra arrangement would be absent. It was also expected that the 6 tetrahedra configuration with pressure averaging would perform better compared to the 6 tetrahedra configuration using individual element pressures due to the reduction in the number of incompressibility constraints [Nagtegaal *et al.*, 1974]. Some test cases were compared to experimental data [Hohler, Schneider, Stilp & Tham, 1978]. Between 37,000 and 39,000 elements were used for these calculations of an $L/D = 10$ tungsten rod striking a 2.5-cm steel plate at normal incidence with a velocity of 1520 m/s. The case of the 6 tetrahedra without pressure averaging produced the closest correlation with experiment for residual velocity and the worst for residual rod length. Furthermore, there was indication of locking in the target grid. The closest correlation for rod residual velocity and residual length was achieved for the symmetric 24 tetrahedra per brick arrangement. However, in all cases, including a 2D calculation, EPIC tended to overpredict residual length by about 16%. This could be due to a number of factors not necessarily related to the grid.

A number of papers present comparisons of plastic strain profiles for Taylor cylinder impacts—the impact of a deformable, short cylinder ($L/D < 3$, generally, although longer rods have been used) striking a rigid surface at impact velocities under 100 m/s. Johnson [1981], for example, compares DYNA, NIKE, and EPIC-2 results with various arrangements of triangles (Fig. 8.5). Many more papers have since appeared to not only test element formulations and arrangements but constitutive relations as well. Such results are interesting but tend to show only minor differences for the various cases considered. It is now clear that the Taylor cylinder is not a very sensitive measure of element arrangement effectiveness; neither is it a good discriminant for testing constitutive models.

c)   *Uniform and variable meshes.*

The effects of mesh size on wave propagation problems can be seen in Figs. 8.6 and 8.7. These depict the impact of a long S-7 tool steel rod ($L/D = 10$, hemispherical nose, $D = 1.02$ cm) into 2.56-cm thick RHA plate at 1103 m/s. The characteristic lengths of this problem are the target thickness and projectile radius. For good results, there should be at least three elements across the radius of the rod. Since uniform mesh spacing is the ideal, this governs the number of elements to be used in the projectile and target.

Figure 8.6 shows initial grids and results at 50 ms after impact. One, three, and five crossed-triangle meshes were used across the projectile radius for the coarse, medium, and fine cases, respectively. The target grid was then set by requiring a 1:1 aspect ratio of all elements in the calculation. The coarse grid computation shows some anomalies near the projectile–target interface and a v-shaped crater, indicative of numerical difficulties with triangular elements. The other two calculations (Fig. 8.6(d) and (e)) appear to be reasonable. The experimentally determined projectile residual mass was 32.1 g, and the residual velocity was 690 m/s. The computed residual masses were 29, 36, and 37 g for the coarse, medium, and fine grids, respectively, while the residual velocities were 600, 670, and 680 m/s. Coarse zoning is adequate for "quick and dirty" scoping calculations or to get preliminary estimates of global quantities such as length loss in the rod, approximate hole size, residual velocity, and overall deformed shapes of the two solids.

290 *Introduction to Hydrocodes*

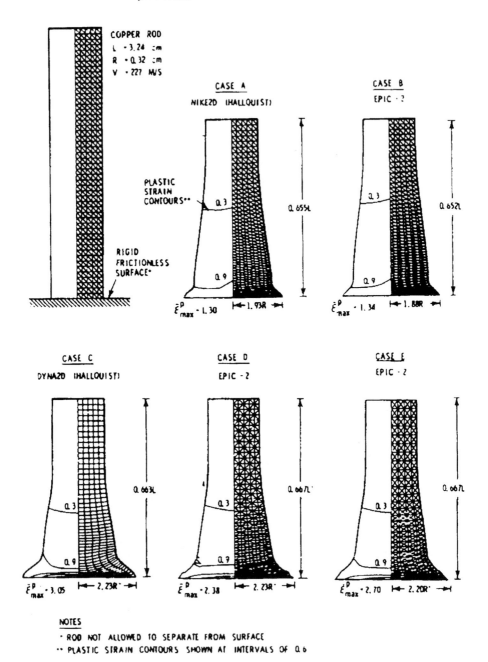

Figure 8.5 Effect of element arrangement on plastic strain distribution [Johnson, 1981].

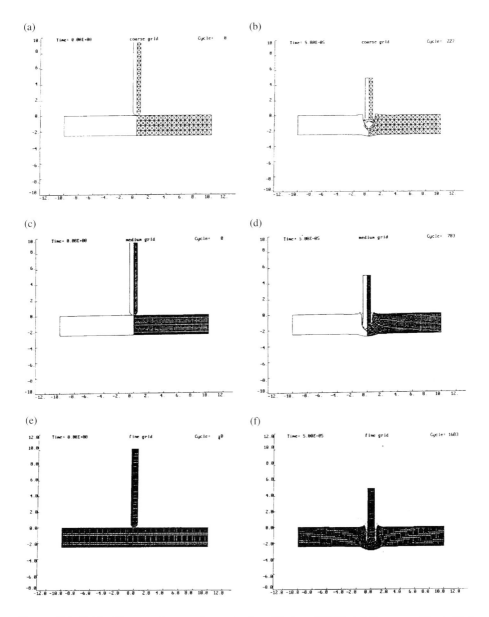

Figure 8.6 Grid effects on long-rod penetration [Zukas, 1993]. (a) Coarse grid - initial. (b) Coarse grid - deformation at 50 μs. (c) Medium grid - initial. (d) Medium grid - deformation at 50 μs. (e) Fine grid - initial. (f) Fine grid - deformation at 50 μs.

It is, however, inadequate to resolve strain and pressure fields with any degree of accuracy. If the calculation employs failure criteria based on stress wave profiles, results from such coarse calculations will be meaningless.

292 *Introduction to Hydrocodes*

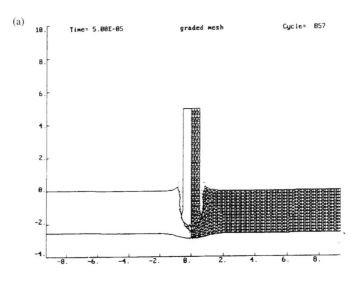

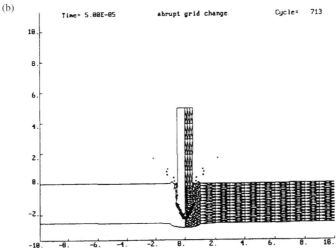

Figure 8.7 Gradual and abrupt mesh change effects. (a) Gradually changing mesh. (b) Abrupt grid changes.

As spatial resolution is increased by a factor of 2 for 2D simulations, computer time rises by a factor of 8 for explicit methods [Zukas & Kimsey, 1991]. Thus, the resolution used should match the accuracy required. If comparison is to be made with time-resolved pressure, stress, or strain data, if internal failure (such as spall) occurs, fine resolution is required for a meaningful calculation, even though the cost is high. Compromising the accuracy of the calculation to save money in these cases is not justifiable since maximum savings come from not doing the calculation at all. This has the added advantage of not generating meaningless numbers that someone not familiar with the events of the calculation might be tempted to believe. On the other hand, if only global (integral) data are available for comparison (residual masses, lengths, hole sizes, deformed shapes), a reasonably crude and inexpensive calculation can be done, provided its interpretation is not pushed too far.

It is possible to take advantage of the localized nature of impact problems when setting up computational grids. Typically, the high strain rates and the high pressures, strains, and temperatures that accompany them are confined to a narrow region, or process zone that extends about three to six striker diameters from the impact interface, depending on striking velocity. Figure 8.7(a) shows results for a calculation that takes advantage of this information by localizing fine zones within three diameters of the impact interface and then gradually expands element size in regions where, at most, elastic waves will propagate. The result is accuracy comparable to the uniformly gridded case but with considerably fewer elements and, therefore, considerable savings in computer time. Care must be taken not to change element sizes too rapidly, however. A change in element size represents a change in element stiffness even if material properties remain the same. Traveling waves, when encountering this stiffness difference, will act as if an impedance mismatch had occurred. Part of the wave will be reflected, part transmitted. If element-to-element size variation is kept below 10%, acceptable results can generally be achieved. Figure 8.7(b) shows the negative aspects of drastic changes in element size.

Using a minimum of three continuum elements across a critical dimension such as a rod radius or plate thickness turns out to be a good rule of thumb for Lagrangian calculations. Eulerian calculations require considerably more [Littlefield & Anderson, 1996]. Some problems, however, such as hypervelocity impact or self-forging fragment formation, may require much more. These are situations where severe pressure gradients exist and move with time. Also situations such as spall, where material failure occurs due to the interaction of stress waves with geometric boundaries, material interfaces, or each other, will require fine resolution. Zukas *et al.* [1993], studying explosively formed penetrator (EFP) formation and penetration, found that five to six elements through the liner thickness and fine zoning in the target were required to match code calculations with experiments. Melosh, Ryan & Asphaug [1992] modeled dynamic fragmentation on a laboratory scale. They developed a fragmentation model and found good correlation with a wealth of experimental data for the largest fragment size and the fragment size–number distribution, provided that an adequate numerical resolution was used. Resolutions of 12×24 cells (where 12 cells defined target radii, which ranged from 2 to 12 cm) were used for most calculations.

Resolutions of 6 × 12 and 24 × 48 cells were also used. All three resolutions gave about the same result. However, substantially finer grids (40 × 80) were needed to match observed near-surface spallation. Johnson et al. [1990] also looked at the effects of grid size on fragment distribution for normal impacts of copper rods at 2 km/s against 1-cm steel plates. Calculations were performed using 1600 (Case A) and 4096 elements (Case B). Figure 8.8 shows the effects of gridding on fragment size distributions for the different grid sizes.

Johnson & Schonhardt [1992] investigated the sensitivity of the EPIC-2 and EPIC-3 codes to a number of factors, including gridding for normal and oblique incidence problems. Normal impact calculations for an $L/D = 10$ tungsten rod striking a 2.5-cm steel plate at a velocity of 1.5 km/s were made with 144, 576, 1296 and 2304 elements. Oblique impact calculations at 60°, measured from the target normal, were made with 480, 1920, 4320 and 7680 elements. With the exception of the lowest resolution, there was relatively little difference for the higher resolution calculations in terms of residual velocity, residual mass, and hole diameter for the 2D calculations. Significant increases in CPU times were observed as resolution was enhanced. Similar results were obtained for the 3D simulations (Fig. 8.9) comparing residual masses, velocities, hole diameters, rotational velocity, and deflection.

Zoning requirements for 2D and 3D Eulerian calculations with the CTH code for long-rod penetration problems are considered in Littlefield & Anderson [1996].

Guidelines for determining resolution and material modeling for practical engineering problems are given in the report of the National Materials Advisory Board Committee on Materials Response to Ultra-High Loading [NMAB, 1980]. The committee recommended an iterative procedure of successive refinements involving computations with existing relatively simple failure descriptions, dynamic material characterization employing relatively simple and standardized techniques, and experimentation to produce useful results for design purposes in many applications. Their report suggests that

> ...rough computations, using simple material models with published or even estimated material properties, may be used in conjunction with exploratory test firings to scope an initial design. Comparison of test data with predictions may reveal discrepancies, which suggest refinements in the computations or material models, and the need for some material property measurements. Once reasonable agreement has been achieved, another round of computations may then be performed to refine the design. Test firings of this design might use more detailed diagnostic instrumentation. This sequence is iterated, including successively more detail in computational models, material property tests, and ordnance test firings, until a satisfactory design is achieved. In this procedure, unnecessarily

Practical Aspects of Numerical Simulations of Dynamic Events 295

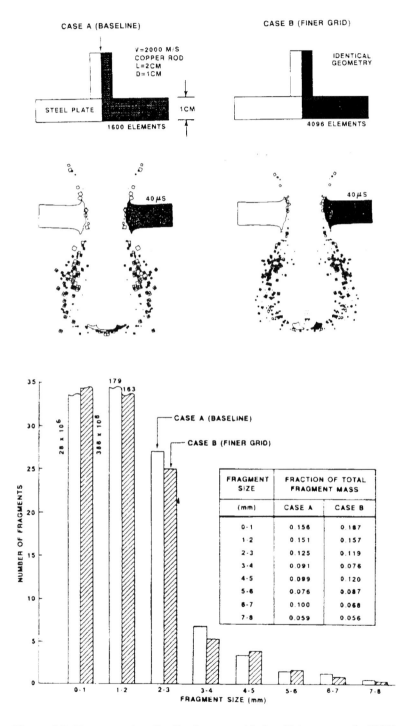

Figure 8.8 Fragment size distribution vs. grid size [Johnson *et al.*, 1990].

296 *Introduction to Hydrocodes*

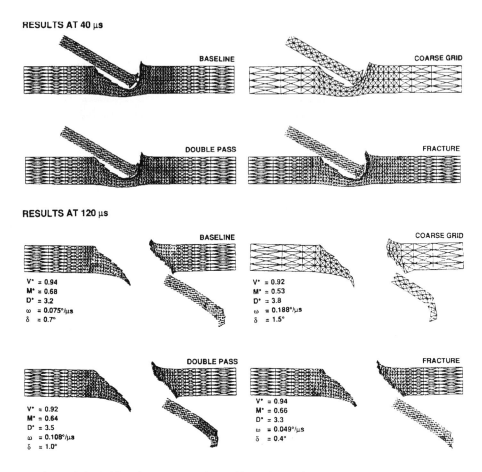

Figure 8.9 Grid and contact surface effects for oblique penetration [Johnson & Schonhardt, 1992].

detailed computations, material property studies or test firings are minimized; only those details necessary to achieve a satisfactory design are included.

d)   ***Abrupt changes in meshes.***

Abrupt grid changes are common in statics calculations. The stress gradient is fixed in space. A grid sufficiently fine for all practical purposes is superimposed over the region. The remainder of the physical body is then modeled with rather large elements to account for the total mass and boundary conditions.

In dynamics calculations, the location of the stress or pressure gradient is a function of time as well as space. It cannot be overemphasized that wave propagation

Practical Aspects of Numerical Simulations of Dynamic Events   297

governs the response. Recall from wave propagation theory that stress waves are reflected from material interfaces, geometric boundaries and each other. Partial reflection also occurs at internal points of discontinuity. Chapter 10, in Fried [1979], describes this very lucidly, as do Bazant and his colleagues [Bazant, 1978; Bazant & Celep, 1982; Celep & Bazant, 1983] for a number of practical systems. A spurious reflection

> ...occurs when the traveling wave crosses over from a fine mesh region to a coarse mesh region. A given mesh size has a lower limit to the wavelength it can approximate or an upper limit to the frequency it can transmit—the cutoff frequency. When the wave enters the coarse mesh region, some of its high-frequency components cannot penetrate and are reflected.... As the wave reaches the larger elements, small waves appear, traveling backwards. Also, because of the largest element size, dispersion becomes more pronounced in the form of leading small waves in front of the original one [Fried, 1979].

In Fried [1979], these points are illustrated with examples of waves in strings. Consider also the following problem, where an $L/D = 3$ mild-steel projectile impacts an armor-steel target at a velocity of 0.8 km/s with sharp mesh discontinuities in both projectile and target. Figure 8.10 shows the grid and results at intermediate times. Figure 8.11 shows the end result. Note that the grid discontinuity has, in effect, pre-determined the outcome. The elements are sufficiently large that the distortions in the finely gridded zone are enhanced by reflected high-frequency waves from the fine/coarse element boundary. Compare the results of Fig. 8.12(a) with those of Fig. 8.12(b), where a graduated mesh was used, and Fig. 8.12(c), with a uniform mesh. Keep in mind that, in high-velocity impact problems, the most severe distortions occur within three to six projectile diameters, depending on impact velocity. To the extent possible, this region should be uniformly zoned, with element size gradually increased by no more than 10% (folklore) from there onward.

8.6   Shortcuts.

When a striker rebounds from the impacted surface of a target, or penetrates along a curved trajectory emerging through the impacted surface with a residual velocity, its behavior is defined as ricochet. Three major factors affect changes in velocity and direction associated with ricochet:

298 *Introduction to Hydrocodes*

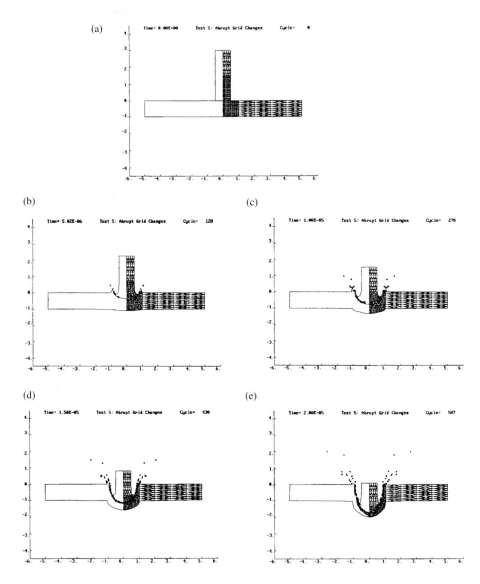

Figure 8.10 Abrupt grid change effects on penetration. (a) 0 ms. (b) 5 μs. (c) 10 μs. (d) 15 μs. (e) 20 μs.

(a) Impact pressures compress and deform striker and target. The subsequent recovery of the stored elastic energy results in motion changes.

(b) The characteristics of the target (surface properties and geometry, material properties, material interfaces) and its subsequent deformation after impact govern

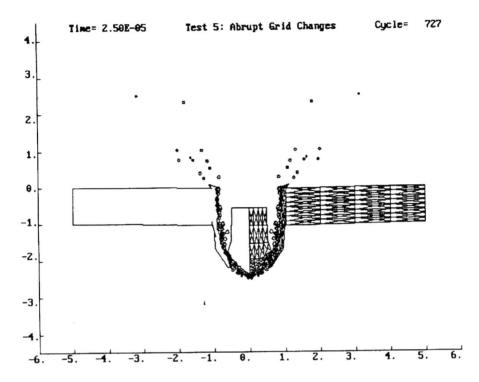

Figure 8.11 Residual penetrator and hole size in the presence of an abrupt grid change.

the direction and magnitude of the resisting force resultant which acts on the striker.

(c) Resistance to motion due to drag and friction reduces velocities.

The component of the resisting force resultant which is aligned with the direction of motion slows the penetrator, which may be a "drag" force. The component of the resultant which acts normal to the direction of motion causes direction change and may be thought of as the "lift" force. If the resisting force resultant does not act on a line which intersects the center of gravity, the striker experiences rotating and bending moments [Recht & Ipson, 1969; Ipson, Recht & Schmeling, 1973].

Ricochet is of interest for a number of reasons. First, there is interest in the basic mechanics of ricochet for rigid and deformable media. Among these is the ingenious work of Johnson, Sengupta & Ghosh [1982a,b] who fired long rods against relatively thick plates, both of Plasticine (modeling clay), at obliquities ranging from 0 to <75° (obliquity is the impact angle measured from the normal to the plate surface). They were

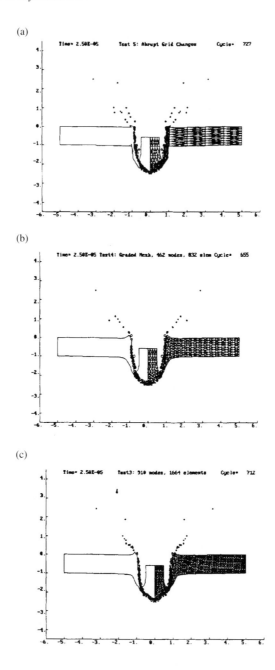

Figure 8.12 (a) Abrupt grid change, (b) gradual change in element size and (c) uniform meshing.

able to ascertain the mechanics of target cratering and define conditions leading to projectile breakup. Many of their observations have since been verified computationally and experimentally for metallic materials. Tate [1979] derived an expression for the ricochet angle (also measured from the plate normal) for long rods from thick plates which agreed well with the observations of Johnson, Sengupta and Ghosh. An extensive experimental program involving ricochet of metallic spheres from metallic targets was undertaken by Backman & Finnegan [1976]. Considerable data on ricochet from sand, clay, water and concrete can be found in the reports by Recht and his colleagues, cited above, as well as Hutchings [1976], Birkhoff [1945], Bushkovitch [1944], Kemper & Jones [1969], Johnson & Daneshi [1979], Daneshi & Johnson [1977] and many others.

The safety of personnel and equipment threatened by ricocheting projectiles or fragments has been an ongoing concern. Typical of such studies are the works of Dunn & Dotson [1964], Hayes, Reeves & Collins [1975], and Reches [1970].

Current interests in ricochet include the areas of hypervelocity impact and forensic engineering. Much of the work in hypervelocity impact involves normal incidence. However, most impacts in reality occur at obliquity. Many space structures in use or in the design phase have irregular outer surfaces or protrusions which are vulnerable to the debris clouds that can be generated by the breakup of projectiles impinging at high obliquity. Schonberg & Taylor [1989a,b] and Schonberg [1990] have experimentally investigated oblique hypervelocity impact and ricochet for aluminum dual-wall structures. They developed a set of empirical equations that characterize observed penetration phenomena as a function of the geometric and material properties of the impacted structure and the diameter, obliquity and velocity of the impacting projectile. Burke & Rowe [1992] review bullet ricochet from the standpoint of forensics. A knowledge of the wounds suffered by shooting victims, the deformation of bullets or shotgun pellets and ricochet marks on surfaces at the scene of a crime can be instrumental in reconstruction of the shooting.

Ricochet is fully a 3D phenomenon. There are pitifully few analytical models which treat oblique incidence. These tend to be either extremely simple or limited to the initial stages of the impact process. Not uncommonly, some empiricism is involved in model development. The ricochet problem can therefore be treated experimentally or with finite element or finite difference codes as a fully 3D solution or with the plane strain approximation.

Experiments take some time to set up, but once materials have been fabricated and the experimental arrangement is in place, 1–4 shots daily may be obtained, depending on whether the experiments are performed at full scale or model scale. Experiments are costly. A typical cost for a single full-scale shot is upwards of $10,000, depending on the complexity of the experimental arrangement, the degree of instrumentation and the data reduction required. Model scale tests in enclosed ranges can be done for about $2,000 per shot. Note that these costs are per shot, nor per data point. Frequently, because of excessive projectile yaw, malfunction of instrumentation or other causes, several shots may be required to obtain a valid data point. Even then, the information extracted from a ballistic test is minimal from the point of view of an analyst—initial and final velocity and orientation of the projectile, residual projectile mass, target deformation and mass loss. Inevitably there is scatter in the data due to variations in material properties of nominally identical materials and uncertainties in initial and boundary conditions. Time and cost constraints almost never permit acquisition of a database with enough variation of parameters to construct unambiguous analytical models.

The cost of 3D simulations is also high, even on supercomputers [Zukas & Kimsey, 1991]. The exact cost depends on the code and computer used, the spatial and temporal resolution, the number of sliding interfaces in the calculation and the constitutive model. However, the cost of a 3D calculation on a supercomputer almost always exceeds the cost of an equivalent experiment. The advantage of the computations is the quantity of information obtained: full time-resolved displacements, strains, strain rates, momenta, energies, forces, moments and so on. Coupled with experimental data, this forms an excellent base for construction of approximate analytical models for parametric studies. The validity of computational results depends on the constitutive model employed, the source of data for the parameters of the constitutive model and the degree of material failure in the experiment and how successfully it is simulated computationally.

In order to avoid the high cost of computing in three dimensions, recourse is sometimes made to 2D calculations employing the plane strain assumption. Two-dimensional plane strain calculations are straightforward enough, relatively inexpensive and provide some useful information about the early stages of impact. However, when the oblique impact of a striker is treated as the impact of an infinitely long wedge (see Fig. 8.13), important physical phenomena are being neglected, not the least of which are the out-of-plane motions leading to lateral stress relaxations. Useful qualitative information (and, if care is taken with the calculations, useful quantitative information) may be obtained from plane strain solutions for the early stages of an oblique impact. Their utility, however, degrades with increasing time after impact so that for late times, when important aspects of penetration and target response (such as bending and shear failure) are being determined, plane strain solutions are speculative at best.

The nature of the plane strain approximation has been discussed by many authors [Bertholf, Kipp & Brown, 1977; Zukas, Jonas & Misey, 1979; Brown, 1982; Zukas, 1982]. The general conclusion of these studies is that, with appropriate scaling, useful information regarding overall kinetic energies, momenta and velocities may be obtained. However, after the time that a release wave would have returned from the lateral boundary of an actual 3D calculation, plane strain calculations are not useful for extracting information related to the internal energy of the problem. They require less

Practical Aspects of Numerical Simulations of Dynamic Events   303

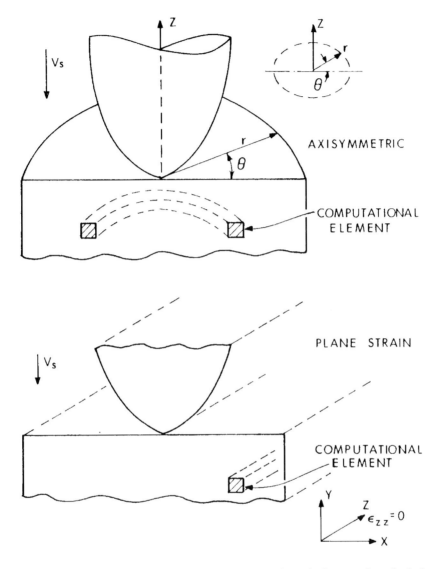

Figure 8.13 Computational elements for axisymmetric and plane strain calculations [Zukas & Gaskill, 1995].

energy for deformation than their exact (2D axisymmetric or 3D oblique impact) counterparts and grossly overestimate late-time deformations.

If they are so fraught with risk, then, why perform plane strain calculations at all? Two reasons stand out: they are cheap and, occasionally, they produce useful results. Norris, Scudder, McMaster & Wilkins [1977] used the HEMP code in plane strain mode to supplement information from model scale oblique impact experiments of long rods

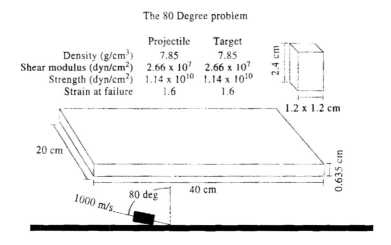

Figure 8.14 Problem setup for 80-degree impact [Zukas & Gaskill, 1995].

striking thin plates at high obliquity and high velocity in order to determine optimum material and geometry configurations for high length-to-diameter (L/D) ratio projectiles. They obtained excellent agreement between calculations and experiments, primarily because penetration was achieved at about the time the reflected wave from the penetrator's lateral dimension arrived. Zukas & Seletes [1991] showed close correlation between experiments and calculations for hypervelocity projectiles striking spaced thin-plate targets at extreme obliquity for deformations and debris spray angles. Other simulations have been less successful. Jonas & Zukas [1978] modeled long-rod ricochet with the EPIC code. Good agreement between plane strain calculations and computed rod deformations was obtained early on after impact. However, comparison of calculations with radiographs at late times after impact showed poor correlation. The plane strain calculations severely underestimated rod bending. Detailed comparisons of the progressive deviation of plane strain calculations from exact (axisymmetric) calculations of long rod impact using the HELP code are shown by Zukas, Jonas & Misey [1979].

Provided some care is taken in setting up calculations, and considerable skepticism employed in interpreting results, plane strain calculations can provide some insights into 3D behavior. For highly energetic problems, the early-time response can be predicted quite well. However, it would be both foolish and dangerous to rely on plane strain calculations alone for design without corroboration by experiments or exact analyses.

As an example, consider the calculation performed by Zukas and Gaskill [1995] of the ricochet of a cylindrical fragment from a thin plate. The impact angle was 80 degrees as measured from the plate normal. Figure 8.14 shows the problem setup.

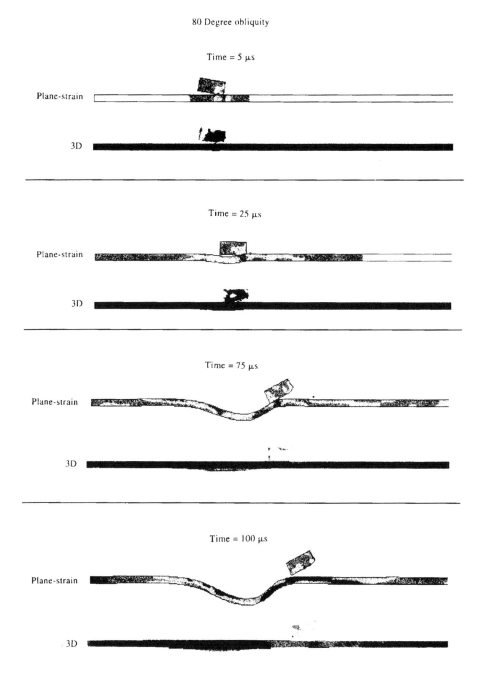

Figure 8.15 Computational results for 2D and 3D computations of projectile ricochet.

Calculations were performed with *ZeuS*, a 2D finite element code for fast, transient loading, and APOLLO, its 3D counterpart, to illustrate the difference between 2D plane strain and fully 3D calculations involving oblique impact and ricochet of compact projectiles. Figure 8.15 shows computational results from both codes for a hardened steel projectile with a mass of 26.5 g (see Zukas and Gaskill [1995] for the geometry, material data and initial conditions used in the calculations) striking a 4130 steel plate at an obliquity of 80 degrees from the plate normal and a velocity of 1 km/s. Experimental data for these situations was taken from the report of Recht & Ipson [1969]. The 3D calculation shows minimal damage to the plate and is in agreement with experimental observations. At early times the 2D plane strain approximation mimics the 3D calculation. At late times, however, once out-of-plane wave effects become important, there is no correlation as regards plate damage. The ricochet angles predicted from 2D plane strain and 3D calculations were 68 and 72 degrees, respectively. Total run time for the 2D calculation on a Silicon Graphics Indigo Extreme workstation was just under 2 min. The 3D calculation required 28 min to reach a problem time of 100 μs.

Plane strain calculations, if used with considerable caution, can produce results which have some qualitative value. However, because of the inherent difference in the physics modeled in the 2D plane strain and 3D calculations, good correlation between the two should not be expected. Plane strain results should never be used for design purposes unless corroborated by exact analyses, experiments or both.

8.7 Summary.

The causes of disagreement between large-scale code calculations and reality for problems involving fast, transient loading have been discussed. The experience of the analyst is a prime factor in the successful use of commercial codes for problems in dynamics. The analyst needs an appropriate educational background and a keen understanding of the physics and mechanics of the problem being addressed. He/she must also have sufficient experience with numerical techniques and large-scale computations in order to select the appropriate computational tool for the problem and to evaluate the computational results. The currently available commercial codes for dynamics problems are in no way "black boxes" that can be assigned to junior engineers with demands for immediate production. The consequences of inappropriate analyses can range from embarrassment and loss of funding through catastrophic failure of poorly designed structures under service loads, with the liabilities and litigations that inevitably follow.

REFERENCES

M.E. Backman and S.A. Finnegan. (1976). Dynamics of the oblique impact and ricochet of nondeforming spheres against thin plates, Naval Weapons Center, Report NWC TP 5844, China lake, CA.

Z.P. Bazant. (1978). Spurious reflection of elastic waves in non-uniform finite element grids. *Comput. Methods Appl. Mech. Engng*, 16, 91-100.

Z.P. Bazant and Z. Celep. (1982). Spurious reflection of elastic waves in nonuniform meshes of constant and linear strain finite elements. *Comput. Struct.*, 15, 451-459.

T. Belytschko and R.J.R. Hughes. (1992). *Computational Methods for Transient Analysis*, New York: Elsevier Science Publishing Co. Third Printing.

L.D. Bertholf, M.E. Kipp and T.W. Brown. (1977). Two-dimensional calculations for the oblique impact of kinetic energy projectiles with a multi-layered target, Ballistic Research Laboratory, Report BRL-CR-333, Aberdeen Proving Ground, MD.

G. Birkhoff. (1945). Ricochet off land surfaces, Ballistic Research Laboratory, Report BRL 535, Aberdeen Proving Ground, MD.

T.W. Brown. (1982). Numerical modeling of oblique hypervelocity impact using two-dimensional plane strain models. In W.J. Nellis, L. Seaman and R.A. Graham, eds, *Shock Waves in Condensed Matter—(1981)*. New York: American Institute of Physics.

T.W. Burke and W.F. Rowe. (1992). Bullet ricochet: a comprehensive review. *J. Forensic Sci., JFSCA*, 37(5), 1254-1260.

A.V. Bushkovitch. (1944). Ricochet of 0.30 cal. Cylinders on striking an earth surface, Ballistic Research Laboratory, Report BRL 499, Aberdeen Proving Ground, MD.

Z. Celep and Z.P. Bazant. (1983). Spurious reflection of elastic waves due to gradually changing finite element size. *Int. J. Num. Methods Engng*, 19(5), 631-646.

B. Creighton. (1984). Numerical resolution calculation for elastic–plastic impact problems, BRL-MR-3418, U.S. Army Ballistic Research Laboratory, Aberdeen Proving Ground, MD.

G.H. Daneshi and W. Johnson. (1977). Forces developed during the ricochet of projectiles of spherical and other shapes. *Int. J. Mech. Sci.*, 19, 661-671.

J. Donea (ed). (1978). *Advanced Structural Dynamics*. London: Applied Science Publishers.

D.J. Dunn and W.D. Dotson, Jr. (1964). A method for estimating danger areas due to ricocheting projectiles, Ballistic Research Laboratory, Report NRL MR 1538, Aberdeen Proving Ground, MD.

I. Fried. (1979). *Numerical Solution of Differential Equations*. New York: Academic Press.

C. Hayes, W. Reeves and J. Collins. (1975). The effects of fragment ricochet on munition lethality, Air Force Armament Laboratory, Report AFATL-TR-75-102, Eglin AFB, FL.

W. Herrman. (1977). Current problems in the finite difference solution of stress waves. *Nonlinear Waves in Solids: Proceedings of the Workshop at University of Illinois,* Chicago, IL, March 21–23.

V. Hohler, E. Schneider, A.J. Stilp and R. Tham. (1978). Length and velocity reduction of high density rods perforating mild steel and armor steel plates. *Proc. 4th Intl Symp. on Ballistics.* Monterey, CA.

I.M. Hutchings. (1976). The ricochet of spheres and cylinders from the surface of water. *Int. J. Mech. Sci.,* 18, 243-247.

T.W. Ipson, R.F. Recht and W.A. Schmeling. (1973). Effect of projectile nose shape on ballistic limit velocity, residual velocity and ricochet obliquity, Naval Weapons Center, Report NWC TP 5607, China Lake, CA.

G.R. Johnson. (1981). Recent developments and analyses associated with the EPIC-2 and EPIC-3 codes. In S.S. Wang and W.J. Renton, eds, Advances in Aerospace Structures and Materials, vol. AD-01. New York: ASME.

W. Johnson and G.H. Daneshi. (1979). Results for the single ricochet of spherical-ended projectiles off sand and clay at up to 400 m/s. In K. Kawata and J. Shioiri, eds, *High Velocity Deformation of Solids.* Berlin: Springer-Verlag.

G.R. Johnson and J.A. Schonhardt. (1992). Some parametric sensitivity analyses for high velocity impact computations. *Nucl. Engng Des.,* 138, 75-91.

W. Johnson, A.K. Sengupta and S.K. Ghosh. (1982a). High velocity oblique impact and ricochet mainly of long rod projectiles: an overview. *Int. J. Mech. Sci.,* 24(7), 425-436.

W. Johnson, A.K. Sengupta and S.K. Ghosh. (1982b). Plasticine modelled high velocity oblique impact and ricochet of long rods. *Int. J. Mech. Sci.,* 24(7), 427-455.

G.R. Johnson, R.A. Stryk, T.J. Holmquist and O.A. Souka. (1990). Recent EPIC code developments for high velocity impact: 3D element arrangements and 2D fragment distributions. *Int. J. Impact Engng,* 10, 281-294.

G.H. Jonas and J.A. Zukas. (1978). Mechanics of penetration: analysis and experiment. *Int. J. Engng Sci.,* 16, 879-903.

W.A. Kemper and J.C. Jones. (1969). Naval Gunfire Study—ricochet from water of 5/38 projectiles, Naval Weapons Laboratory, Report NWL TR 2252, Dahlgren, VA.

D.L. Littlefield and C.E. Anderson Jr. (1996). A study of zoning requirements for 2-D and 3-D long-rod penetration. In S.C. Schmidt and W.C. Tao, eds, *APS Topical Conference on Shock Compression in Condensed Matter,* AIP Conf. Proc. 370, 1131-1134, Melville, New York. American Institute of Physics.

R.H. MacNeal. (1994). *Finite Elements: Their Design and Performance.* New York: Marcel Dekker.

R.J. Melosh and S. Utku. (1983). Principles for design of finite element meshes. In A.K. Noor and W.D. Pilkey, eds, *State-of-the-Art Surveys on Finite Element Technology*. New York: ASME.

H.J. Melosh, E.V. Ryan and E. Asphaug. (1992). Dynamic fragmentation in impacts: hydrocode simulation in laboratory impacts. *J. Geophys. Res.*, 97(E9), 14735-14759.

C. Meyer (ed). (1989). *Finite Element Idealization*. New York: ASCE.

A.J. Morris and R. Vignjevic. (1997). Consistent finite element structural analysis and error control. *Comput. Methods Appl. Mech. Engng*, 140, 87-108.

J.C. Nagtegaal, D.M. Parks and J.R. Rice. (1974). On numerically accurate finite element solutions in the fully plastic range. *Comput. Methods Appl. Mech. Engng*, 4, 153-178.

National Materials Advisory Board. (1980). Materials response to ultra-high loading rates, NMAB-356, Washington, DC.

D.M. Norris, K.J. Scudder, W.A. McMaster and M.L. Wilkins. (1977). Mechanics of long rod penetration at high obliquity, in *Proc. High Density Alloy Penetrator Materials Conf.*, Army materials and Mechanics Research Center, publication AMMRC-SP-77-3, Watertown, MA.

M. Reches. (1970). Fragment ricochet off homogeneous soils and its effects on weapons lethality, Army Material Systems Analysis Agency, Report AMSAA TM 79, Aberdeen Proving Ground, MD.

R.F. Recht and T.W. Ipson. (1969). The dynamics if terminal ballistics, Denver Research Institute, Final Report, Contract #DA-23-072-ORD-1302 (AD 274128), Denver, CO.

W.P. Schonberg. (1990). Hypervelocity impact penetration phenomena in aluminum space structures. *ASCE, J. Aerospace Engng*, 3(3).

W.P. Schonberg and R.A. Taylor. (1989a). Oblique hypervelocity impact response of dual-sheet structures, National Aeronautics and Space Administration, Report NASA-TM-100358.

W.P. Schonberg and R.A. Taylor. (1989b). Penetration and ricochet phenomena in oblique hypervelocity impact. *AIAA J.*, 27(5).

R. Skalak. (1957). Longitudinal impact of a semi-infinite circular elastic bar. *J. Appl. Mech., Trans. ASME*, 24, 59-63.

B.A. Szabo and R.L. Actis. (1996). Finite element analysis in professional practice. *Comput. Methods Appl. Mech. Engng*, 133, 209-228.

A. Tate. (1979). A simple estimate of the minimum target obliquity required for the ricochet of a high speed long rod projectile. *J. Phys. D: Appl. Phys.*, 12, 1825-1829.

S. Utku and R.J. Melosh. (1983). Solution errors in finite element analysis. In A.K. Noor and W.D. Pilkey, eds, *State-of-the-Art Surveys on Finite Element Technology*. New York: ASME.

O.C. Zienkiewicz. (1983). *Finite Elements in the Time Domain*. In A.K. Noor and W.D. Pilkey, eds, *State-of-the-Art Surveys on Finite Element Technology*, NY: ASME.

J.A. Zukas. (1979). Some problems with the EPIC-2 code, Internal Memorandum IMR 647, U.S. Army Ballistic Research Laboratory, Aberdeen Proving Ground, MD.

J.A. Zukas. (1982). Penetration and perforation of solids. In J.A. Zukas, T. Nicholas, H.F. Swift, L.B. Greszczuk and D.R. Curran, eds, *Impact Dynamics*, reprinted 1992 by Kriger Publishing Co., Malabar, FL.

J.A. Zukas. (1993). Some common problems in the numerical modeling of impact phenomena. *Comput. Syst. Engng*, 4(1), 43-58.

J.A. Zukas and B. Gaskill. (1995). Ricochet of deforming projectiles from deforming plate. In Y.S. Shin, ed, *Structures Under Extreme Loading Conditions*, PVP-Vol. 229. New York: ASME.

J.A. Zukas and K.D. Kimsey. (1991). Supercomputing and computational penetration mechanics. In R.F. Kulak and L.E. Schwer, eds, *Computational Aspects of Contact, Impact and Penetration*. Lausanne, Switzerland: Elmepress International.

J.A. Zukas and D.R. Scheffler. (2000). Practical aspects of numerical simulation of dynamic events: effects of meshing, Army Research Laboratory, ARL-TR-2303.

J.A. Zukas and S.B. Seletes. (1991). Hypervelocity impact on space structures. In T.L. Geers and Y.S. Shin, eds, *Dynamic Response of Structures to High-Energy Excitations*, AMD-vol. 127/PVP-vol. 255. New York: ASME.

J.A. Zukas, G.H. Jonas and J.J. Misey. (1979). On the utility of the plane strain approximation for oblique impact computations, Ballistic Research Laboratory, Report BRL-MR-02969, Aberdeen Proving Ground, MD.

J.A. Zukas, C.A. Weickert and P.J. Gallagher. (1993). Numerical simulation of penetration by explosively-formed projectiles. *Propellants, Explosives and Pyrotechnics*, 18, 259-263.

SUBJECT INDEX

Artificial viscosity 138–141, 281, 284
   tensor viscosity 140
Arbitrary Lagrange-Euler (ALE) 15, 123, 201, 223–226

Ballistic limit 22
Boundaries 40, 49–51, 55–58, 62, 66–70, 115–117, 180, 182–183, 173, 227

Conservation of mass 86, 107, 201, 208
Conservation of momentum 86, 107, 201, 208
Conservation of energy 86, 107, 190, 201, 208
   Constitutive model 1, 106, 108–109, 148–160, 251–279
   Bodner-Partom 152–155
   CAP 158–160
   incremental-elastic-plastic 150–151
   porous materials 156–157
   Steinberg 153
   Zerilli-Armstrong 151–152
Constitutive relationship see Constitutive model
Coupled Euler-Lagrange methods 123, 201, 226–235

Deformation 14–15
Displacement 174–176, 256

Equation of State (EOS) 91, 93–101, 110, 189–190, 194
   Becker-Kistiakowski-Wilson (BKW) 98, 110
   ideal gas 110
   Jones-Wilkins-Lee (JWL) 99, 110
   Los Alamos 110, 190
   Mie-Gruneisen 94–95, 109, 189–190, 194
   SESAME 189
   Tillotson 95–97, 109, 189–190

Euler methods 123, 125, 201, 204–233
   grids 211
   material transport 201–202, 208, 210–223
   material interface treatments 210–223
      BLINT 217, 220
      HULL diffusion limiter 210
      SLIC 214–216
      SMYRA 217–218
      Youngs algorithm 214–218
   meshes 123–125
Experimental methods 103, 251–275
   expanding ring test 265–268
   plate impact test 268–271
   pressure-shear test 272–274
   split-Hopkinson pressure bar test 252–260, 272
   Taylor cylinder test 260–265
Explosives
   energy 193–195
   equation of state 98–99, 109–110
   loading 66–67
   numerical processing 193–195
   phenomena 148–160
   pressure 193–195

Finite difference 106, 111–117, 141
Finite element 106, 117–122

High strain rate phenomena 251–275
Hugoniot
   curves 81, 87–92, 100
   Elastic Limit 81–82
   jump conditions 107–108
   off-Hugoniot 268
   plate impact tests 268–269
Hydrocodes 22, 75, 161, 169–198, 279–307
   APPOLO 305–306

AUTODYN 136, 148, 169
CTH 189, 217, 222
DORF 110
DYNA 15, 110, 126, 135–136, 148, 169, 174, 177, 284
DYSMAS 127, 132, 136, 148, 169, 221, 226, 284
EPIC 110, 126, 136–138, 156, 169, 174–175, 177, 189, 193, 195, 197, 286, 290, 294–296
HELP 1110, 126, 204, 206–209, 211–213
HEMP 110, 126, 187, 193
HULL 110, 126, 156, 210–211, 227–236
KRAKEN 222
MAGI 242
MESA 222
OIL 110, 187, 204, 206
PIC 206
PRONTO 136, 148, 169
RPM 206
TENSOR 156
TOODY 110
ZeuS 38, 129, 131, 133–137, 169–170, 174–175, 177–185, 189–190, 192–193, 196, 203–205, 291–292, 298–300

Lagrangian methods
  alternatives 201–245
  arbitrary Euler-Lagrange 201, 223–226
  artificial viscosity 138–139
  contact-impact 126–133, 176–185
    contact surfaces 129–130, 173
    multiplier method 128
    penalty method 128
    Pinball algorithm 132–134
    put-back logic 128–129
  large distortions 134–138
  meshes 105–141, 123–138, 170–174
    abrupt mesh changes 292–293, 296–300
    element aspect ratio 285–286
    element arrangement 287–288

uniform and variable meshes 289–295
nodal forces 195–198
Smoothed Particle Hydrodynamics (SPH) 123, 201, 237–238
Loading/response times 12, 22

Material failure 14, 24, 61, 66, 106, 192–193, 270–271, 280
Material models see Constitutive equations

Oblique impact 66, 138, 297–306

Plane strain solutions 297–306
Plasticity
  effective (equivalent) plastic strain 40, 192
  incremental elastic-plastic 206
  radial return algorithm 187, 191
  von Mises yield function 188
  wave motion 70, 82–83
Pressure 1, 14, 79, 83–92, 99, 172, 189, 251

Rankine-Hugoniot jump conditions 85–86
Ricochet 299–306

Smoothed-Particle Hydrodynamics (SPH) 235–245
  B-spline 240–241
  kernel estimates 240–242
Sources of computational error
  mesh 285–297
    abrupt mesh changes 296–297
    aspect ratio 285–287
    element arrangement 287–289
    recommended procedure 294–295
    variable mesh size 289–294
  human 284
Stress 1, 36, 46, 48, 58–63, 68, 78–83, 172, 186, 207, 209, 238–239
  deviators 108–109, 207, 209, 238–239
  finite rise times 48
  intensity 47–48
  layered media 62–63
  split-Hopkinson pressure bar 257

Subject Index 313

Taylor cylinder test 260
Uniaxial 34, 71, 75
wave propagation 34–6, 40, 43, 57–60, 67–70
Stress-strain curve 35, 75–77, 80, 257–258, 260, 261, 268, 272, 274
Strain 1, 14, 23, 78–83, 118, 120–122, 151–152, 172, 185–186, 256–257
uniaxial 34, 71, 75–82
Strain rate 1, 12–13, 22, 33, 172, 185–186, 279
Structural dynamics 1, 4–16, 26
aerospace structures 10
blast loading 4–6, 149–150
codes
ABAQUS 127
ADINA 148
MAGNA 148
MARC 148
NASTRAN 148
NIKE 148
containment structures 8
fluid-structure interaction 6
jet cutting technology 6–7
liquid–solid impact 6
mechanical systems 8
non-perforating impacts 11
vehicle collisions 9–10

Time integration 106, 141–148
explicit 142, 146–146, 281
implicit 146–148

Velocity 1, 41–49, 55–57, 86–100, 108–109, 174–176
particle 44–46, 48–49, 86–87, 89–92, 99
strains 108–109
striker wave motion 41, 44–46
wave reflections 55–57

Wave propagation 1, 16–26, 33–72, 251
applications 15–17
explosive hardening, welding, forming 15, 17
lunar & planetary impact 16
shock consolidation 17
bending (flexural) waves 38, 55–56
in rods and plates 34, 36–46
interfacial (Stoneleigh) waves 38
longitudinal waves 34, 37, 40, 50–53
plastic waves 70
shear waves 34, 38
shock velocity 86–91
shock waves 75–101, 122
sound speed 44–45, 47, 49, 50–53, 82–83, 191
surface waves 38
torsional waves 34, 54–55
wave equation 3–4
normal mode solution 3
traveling wave solution 3
waves in layered media (Love waves) 38, 60, 64–66